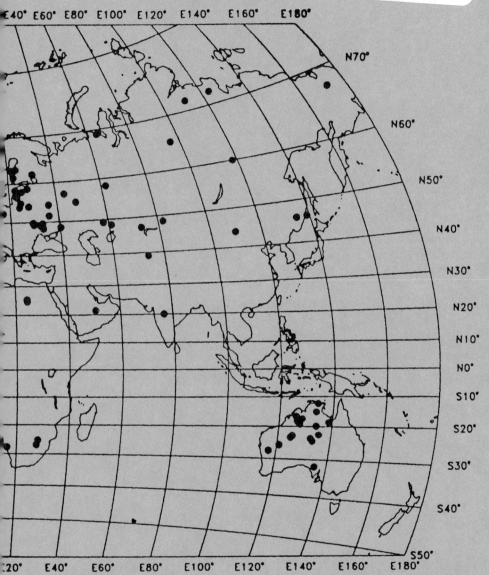

# The locations of asteroid craters on the Earth
Art courtesy of R. A. F. Grieve, Geological Survey of Canada

# DOOMSDAY ASTEROID

# Other books by Donald Cox

*The Space Race: From Sputnik to Apollo and Beyond*

*Spacepower* (with Michael Stoiko)

*Islands in Space: The Challenge of the Asteroids*
(with Dandridge Cole)

*Rocketry Thru the Ages* (with Michael Stoiko)

*You in the Universe* (with Michael Stoiko)

*America's New Policy Makers: The Scientists' Rise to Power*

*American Explorers of Space*

*The Perils of Peace: Conversion to What?*

*The City as a Schoolhouse*

*Pioneers of American Ecology*

*Stations in Space*

*Explorers of the Deep*

*Mafia Wipeout: How the Law Put Away an Entire Crime Family*

*Hemlock's Cup: The Struggle for Death with Dignity*

*The Joseph Priestley Science Center (Ford Foundation)*

# DOOMSDAY ASTEROID

## Can We Survive?

**DONALD W. COX, ED. D.
& JAMES H. CHESTEK, P.E.**

Prometheus Books

59 John Glenn Drive
Amherst, New York 14228-2197

Excerpt from *Rendezvous with Rama,* copyright © 1973 by Arthur C. Clarke, reprinted by permission of Harcourt Brace & Company.

Published 1996 by Prometheus Books

00 99 98 97     5 4 3 2

Library of Congress Cataloging-in-Publication Data

Cox, Donald W. (Donald William), 1921–
    Doomsday asteroid : can we survive? / by Donald Cox and James H. Chestek.
        p.    cm.
    Includes bibliographical references and index.
    ISBN 1–57392–066–5 (cloth : alk. paper)
    1. Asteroids—Collision with Earth. 2. Near Earth asteroids. I. Chestek, James H. II. Title.
QB377.C69    1996
363.3'49—dc20                                                                96–27159
                                                                                    CIP

Printed in the United States of America on acid-free paper

# Contents

## Part Two: The Search and Landing Process

## Part Three: Colonizing the Minor Planets

## Part Four: Cosmic Cooperation Is Possible

# Foreword

# The New Frontier

This is a book about the danger, challenge, and promise of a new frontier.

History shows that a nation is only as good as the challenges it is willing to accept. For the first four centuries of our history, Americans were faced with, and enthusiastically embraced, the challenge presented by our open western frontier. In its harsh but free environment, Americans were given room to innovate socially and technologically. Practicality overruled custom, invention was welcomed, and progress—the ability of human beings to change the world as they found it for the better—was celebrated. Thus was born the classic "can-do" American character.

In the twentieth century, however, the western frontier has evaporated, and with its passing our society has become threatened with having its fascination with innovation and progress replaced by bureaucracy and stasis. We would not be the first civilization to travel that road. In societies, as in people, without effort muscle turns into fat, and enthusiasm wanes.

But as John F. Kennedy pointed out in leading America's first push into space, there is a new and vast frontier waiting for us just a few hundred miles above our heads.

With respect to both its challenge and its extent, the frontier of space dwarfs any that humanity has faced in the past. The moon is a world with

11

a surface area equal to that of Africa, Mars is a planet equal in area to all the continents of Earth combined. Beyond Mars is an asteroid belt, filled with tens of thousands of miniature worlds, each with areas the size of small countries; then four giant outer planets, hundreds of times the size of the Earth, accompanied by dozens of moons, six of which are larger than Eurasia. And our solar system is just the beginning; in our own galaxy there are likely at least a hundred billion more such systems.

Space is harsh, but it also has vast riches. In space, solar energy is available twenty-four hours a day without obstruction by clouds. With our first tentative explorations of the moon we have already found an element there, helium-3, which does not exist on Earth but which could be invaluable as a fuel in future pollution-free fusion reactors. On Mars we have found, in addition to every element required to support human life and civilization, deuterium, worth $10,000 per kilogram on Earth, in quantities five times greater than on our home planet. In this book, Donald Cox and James Chestek make a strong case for the economic potential of the huge sources of wealth in metals and other resources known to exist among the asteroids. Without doubt in the vast reaches of these worlds and others, there is much more waiting to be discovered.

But like the New World revealed to European civilization by the voyage of Columbus, the new worlds of space have a value that is much greater than any treasure that they can provide. They can become homes to new, diverse, and dynamic branches of human civilization.

Those who view the potential for extraterrestrial colonization with skepticism due to its apparent difficulties generally fail to comprehend how difficult the colonization of North America was in its day. The challenge of the wilderness continent was enormous—indeed, half the pilgrims died in their first year here—but the pioneers took it on with grit and ingenuity, becoming better people, and inventing a better civilization than any that had existed before in the process. Similarly, the pioneers of the new worlds in space will be forced to develop means for using the resources of alien lands to produce all the necessities of life. They will be forced to invent new technologies and, as in the American past, will be faced with a severe labor shortage that will incite them to find better ways to develop and use the potential of every citizen. In the nineteenth century, these imperatives drove "Yankee ingenuity" to produce the waves of invention and innovations in education and mass production that revolutionized life, not only in America, but across the globe. In the twenty-first century, as I point out in my book, *The Case for Mars*, it may be hothouse inventor culture and ingenuity forced upon a Martian colony which, in the course of solving its own vital problems, develops the breakthroughs in biotechnology, robotics, hydroponic greenhouse agriculture, energy production, and social organization that will greatly advance the human condition on Earth. If so, then the income from licensing such inventions on Earth could also provide all the cash the Martian settlers need to pay for required terrestrial imports as well.

Necessity is the mother of invention: the worlds of the new frontier can provide the cradle.

## Rogue Asteroids and Doomsday Comets

But this book is focused more on rogue asteroids and doomsday comets; you know, big rocks impacting the Earth and wiping out dinosaurs and unwary human civilizations. Why then, in introducing this book, do I spend so much time talking about the promise of the space frontier? The answer is that while averting doomsday asteroids is an important subject, it is ultimately merely a subset of a much more important endeavor, that of conquering the space frontier. It is the fact that the authors, space-science writer Don Cox and aerospace veteran Jim Chestek, appreciate this, that makes this book worthwhile. Indeed, it is the absence of this insight that reduces most other books on the asteroid menace to mere fashionable alarmism.

Life on Earth has survived and prospered because at an early date it was able to take control, dictating the physical and chemical conditions of the planet in defiance of both solar and geological cycles. If it had not done this, terrestrial life would long since have gone extinct, as the sun today is more than 40 percent brighter than it was at the time of life's origin. Without the ability of that life to control terrestrial temperatures by regulating levels of $CO_2$ and the content of other greenhouse gases in the atmosphere, our ancestors would have all been cooked billions of years ago. Moreover, by replacing the $CO_2$ content of the atmosphere with oxygen, plant life on Earth transformed the terrestrial chemical environment to favor the development of species with the capability for increased activity and intelligence, and ever more rapid evolution of still higher and more complex forms.

Within the history of the biosphere itself, the same phenomenon repeats. Those groups of species—whether natural ecosystems or human civilizations—whose activities effectively control their surrounding environments to favor their own growth are those that survive, those which do not risk extinction. In the game of life, the only way to win is to have a part in making the rules.

On the short time scale, the relevant environment for most species is the Earth, and most ecosystems can get by for a fair while if they can deal with developments below the stratosphere. But over the long haul this is not true. As Cox and Chestek make clear, since the Alvarez hypothesis successfully explained the mass extinctions that occurred at the end of the age of dinosaurs as having been caused by asteroidal bombardment, it is now apparent that the relevant environment for life on Earth is not merely the planet of residence, but the whole solar system.

Few people today understand this, yet it is true that subtle events in the asteroid belt or the Oort cloud (a cloud of small bodies postulated to orbit the sun beyond the orbit of Pluto and act as a reservoir of comets) determined the fates of their ancestors, and may well in the future determine the fates of their descendants. It seems so unbelievable that invisible happenings so far away could matter so much here. Similarly, throughout history, few inhabitants of rustic villages going about their daily lives were aware of the machinations of politicians and diplomats in their nation's capitals, which periodically would sweep the villagers off to die in cataclysmic wars on distant battlefields.

What you can't see *can* kill you. What you can't control probably will.

Humanity's home, humanity's environment, is not the Earth, it is the solar system. We've done well for ourselves so far by taking over the Earth and changing it in our own interests. Most people today, at least in the world's more advanced sectors, can walk about without fear of being dismembered by giant cats, are assured of sufficient food and fuel to survive the next winter, and can even drink water without risk of death. Sabretooth tigers, locusts, and bacteria have, for now at least, been beaten. But in a larger sense we're still helpless. We may feel safe, having thrown our village's hoodlums in jail, but in the capital, behind the scenes, diplomats are meeting with generals, and plans are being made. . . .

Our environment is the solar system, and we won't control our fate until we control it. The geological record is clear. Asteroids do hit. Mass extinctions of species every bit as dominant on Earth in their day as humanity is in ours do occur.

We can't shoot down an incoming asteroid with an antiaircraft gun, air to air missile, or Star Wars defense system. If we want to prevent asteroid impacts, we have to be able to direct the course of these massive objects while they are still hundreds of millions of miles away from the Earth.

In short, if humanity wants to either progress or survive, we have to become a spacefaring species. That is what this book is about. Read it and enjoy.

In the end, it is creativity, not austerity, that will be the key to our survival.

Dr. Robert Zubrin
May 1996

Dr. Robert Zubrin is a distinguished aerospace scientist and engineer. His first claim to prominence was his proposal for a "Mars Direct" mission which would provide a thrust for the colonization of Mars using current technology, but exploiting the carbon dioxide in the Mars atmosphere to provide propellants for the return rockets. Although this idea has never been adopted by NASA, it has been widely endorsed as a breakthrough concept; that of "living off the land" as an essential ingredient of moving into the solar system in a significant way.

Dr. Zubrin is a Director of the National Space Society, and has been elected as chairman of its Executive Committee. He is also a Fellow of the British Interplanetary Society. He is also the author of *The Case for Mars* and *Islands in the Sky*, co-edited with Stanley Schmidt. He was a senior Staff Engineer for the Lockheed-Martin Company who now heads his own company, Pioneer Astronautics, in Denver, Colorado.

# Prologue

# Spaceguard

Sooner or later, it was bound to happen. On June 30, 1908, Moscow escaped destruction by three hours and four thousand kilometers—a margin invisibly small by the standards of the universe. On February 12, 1947, another Russian city had a still narrower escape, when the second great meteorite of the twentieth century detonated less than four hundred kilometers from Vladivostok, with an explosion rivaling that of the newly invented uranium bomb.

In those days there was nothing that men could do to protect themselves against the last random shots in the cosmic bombardment that had once scarred the face of the Moon. The meteorites of 1908 and 1947 had struck uninhabited wilderness; but by the end of the twenty-first century there was no region left on Earth that could be safely used for celestial target practice. The human race had spread from pole to pole. And so, inevitably . . .

At 0946 GMT on the morning of September 11 in the exceptionally beautiful summer of the year 2077, most of the inhabitants of Europe saw a dazzling fireball appear in the eastern sky. Within seconds it was brighter than the Sun, and as it moved across the heavens—at first in utter silence—it left behind it a churning column of dust and smoke.

Somewhere above Austria it began to disintegrate, producing a series

17

of concussions so violent that more than a million people had their hearing permanently damaged. They were the lucky ones.

Moving at fifty kilometers a second, a thousand tons of rock and metal impacted on the plains of northern Italy, destroying in a few flaming moments the labor of centuries. The cities of Padua and Verona were wiped from the face of the Earth; and the last glories of Venice sank forever beneath the sea as the waters of the Adriatic came thundering landward after the hammer blow from space.

Six hundred thousand people died, and the total damage was more than a trillion dollars. But the loss to art, to history, to science—to the whole human race, for the rest of time—was beyond all computation. It was as if a great war had been fought and lost in a single morning; and few could draw much pleasure from the fact that, as the dust of destruction slowly settled, for months the whole world witnessed the most splendid dawns and sunsets since Krakatoa.[*]

After the initial shock, mankind reacted with a determination and a unity that no earlier age could have shown. Such a disaster, it was realized, might not occur again for a thousand years—but it might occur tomorrow. And the next time, the consequences could be even worse.

Very well; *there would be no next time.*

A hundred years earlier, a much poorer world, with far feebler resources, had squandered its wealth attempting to destroy weapons launched, suicidally, by mankind against itself. The effort had never been successful, but the skills acquired then had not been forgotten. Now they could be used for a far nobler purpose, and on an infinitely vaster stage. No meteorite large enough to cause catastrophe would ever again be allowed to breach the defenses of Earth.

So began Project *Spaceguard*. Fifty years later—and in a way that none of its designers could ever have anticipated—it justified its existence.

---

*Krakatoa is a volcanic island between Java and Sumatra (in the Indian Ocean) that erupted in 1883, causing a tsunami (tidal wave) that killed 36,000 people.

# Preface

It is difficult to say what is impossible. The dream of yesterday is the hope of today, and the reality of tomorrow.
—Robert H. Goddard (1882–1945),
"Father of American Rocketry"

The human race, and particularly that segment of the race living in the United States, is traveling down the road to the future at high speed. The path we follow is that of technological change and potential disaster.

While we have little hope of stopping or slowing the technological race, we can learn to improve our foresight enough to be better prepared for the changes as they take place. Some instruction in forecasting and the nature of the probable future might even be included in our school curricula. While we know nothing for certain about the future, we only know little for certain about the past. And yet, we spend considerable time studying the probable past. It would seem reasonable to spend a little time studying the probable future. After all, it is the future in which we will spend our lives, not the past.

Of course, no one can say for certain what will happen in the future, just as no one can say for sure what is possible and impossible. But we can make carefully reasoned estimates about what we believe to be possible and, more often than not, come close to the mark.

Humanity has greatly altered the biosphere over the past one hundred years—including major pollutions of the land, water, and the atmosphere to the point that the spectre of an imminent global catastrophe confronts us on all three of the above frontiers. It is from another realm, however, that we face the possible immediate extinction of humanity: namely, the threat of a cosmic outer space invader striking the Earth sometime soon.

While global warming and the collection of greenhouse gases can radically alter the Earth's climate and atmosphere to pose a long-term threat to the survival of *Homo sapiens*, it is worth noting that in the last 600 million years, the Earth has experienced five major extinctions of life, all usually linked to severe climatic changes.

In each of these five catastrophic events, between 35 and 95 percent of all the then living species on this fragile planet disappeared. Scientists know from fossil records that in each case it took 10 million years or more for the Earth to come back and regain its biological diversity.

Biologists fear that humanity may now be precipitating a self-made extinction from *within* on a comparable scale. But it is from *without* that humankind faces an uncertain future, *when* and *if* a giant comet or an asteroid comes crashing down somewhere on Earth.

We know about the recent extinctions of thousands of diverse species of flora and fauna, of fish, birds, and animals, wrought by our tinkering with the biosphere. What we don't know is the dangers from the cosmos (which is what this book is all about).

As humankind starts to grapple with the consequences of its own folly (which was examined in depth at the United Nations-sponsored conference in Rio de Janiero on the global environment in June 1992), it is also time to examine the prospects of a stray asteroid plunging into our planet in the near future.

Where such a hit will come and when, we do not know. But we do know the potential exists for such a cataclysm, wrought by an outer space-borne doomsday rock—wreaking the megaton power of hundreds of H-bombs all going off at once. Such a hit can bring us the deep chill and darkness of a cosmic nuclear winter—and can wipe out humanity as we know it.

Before we can agree on possible methods of staving off such a planetary disaster, we first must attain a basic understanding of the nature of the threat we face.

# A Spark Strikes in the Hair of the Tiger: A Personal Observation by Don Cox

I was sitting in front of a giant fireplace on a cold, blustery mid-April 1992 Sunday afternoon in the Ship Room of the historic Nassau Inn at Princeton, New Jersey, with some crackling logs as the only sounds breaking the stillness. While there I contemplated the meaning of the just-released House Science Committee's *Spaceguard Survey*, a report on the threat of Near-Earth Objects (NEOs) to our civilization. Ironically, I had coauthored a book on the prospects of future spaceflight to the asteroids (which are also known as Near-Earth Objects) some thirty-one years before with the late Princeton-educated space scientist, Dandridge Cole. (Cole was our *first* American expert on the importance of these strange pieces of what was probably once a giant planet located between Mars and Jupiter that broke apart eons ago at the dawn of the birth of our solar system.)

I looked up at the walls around the cozy room and saw framed prints of now extinct nineteenth-century sailing ships that once prowled the seven seas. I couldn't help but make the connection in my mind between the atmosphere of this hospitable, nautical room in an old college town and the title of our long ago, out-of-print book, *Islands in Space: The Challenge of the Planetoids\**—the first major scientific history of the asteroids.

The thought occurred to me then, "why not update and revise that joint effort?" since no other similar book had been published in the interim. I also remembered the sage advice of the Nobel Prize-winning physicist Leon Lederman, of the University of Chicago, who wrote recently that "there is a crying need for a popular understanding of modern science in our culture."

With that thought in mind, I opted to try to rewrite our first attempt to throw some light on a dark corner of the cosmos, to provide some understanding for the layperson of the thousands of previously ignored tiny asteroids that are whirling about in our solar system. This mission became imperative in the light of the National Aeronautics and Space Administration's (NASA) April 1992 recommendation to Congress to allocate $25

---

\*(Radnor, Penn.: Chilton, 1964).

million immediately for a thorough search and tracking of these Near-Earth Objects which might be threatening the Earth with a cataclysm similar to the one that annihilated the dinosaurs 65 million years ago.

In the course of working on the book, I encountered some obstacles and some good fortune. I discovered that major parts of the book were seriously outdated in technical content, and I knew I lacked the scientific expertise to update this material. At a meeting of a local chapter of the National Space Society, I met a man who *did* have that ability. As we talked, I discovered that he had been a colleague of Dan Cole's during the writing of *Islands in Space*, and in fact, had read that book in manuscript and offered suggestions to Cole.

It turns out that he had also considered a revision to *Islands*, but had not pursued the idea because he did not know who held the copyrights and so forth. We agreed to work together, and I believe that we have a better book as a result of our joint efforts.

## Why Asteroids Are Important to Us

Before the beginning of the Space Age, in 1957, when *Sputnik I* ascended into orbit, the planetoids were almost completely neglected. In fact, many astronomers regarded these "minor planets" as no more than nuisances, and, shortly before *Sputnik I,* it was seriously suggested that even the small amount of continuing work in tracking and cataloging them be abandoned. (As a further indication of this lack, only a single slim book which treats the subject of planetoids exclusively was available prior to the publication of *Islands in Space.*[1])

Our book tells the updated story of the planetoids, the three-dimensional islands of the new three-dimensional cosmic sea; it shows why they are of great interest to pure (theoretical) or basic science; why they can be explored more easily than any of our other neighbors in space; why the return-on-investment rationale promises to be greater than that for other space ventures; why they can be of tremendous value for colonization and as sites in the exploration of the rest of the solar system; and why they may be of military and commercial as well as scientific interest.

The great debate on the purpose and value of spaceflight which is destined to continue into the twenty-first century will turn the attention of scientist and layperson alike toward the continents and islands of the new sea. What can we gain here? What can we learn? What can we accomplish? What could be our future in space if we dare to take up the challenge?

While thoughtful people throughout the world are coming to recognize that the twentieth century is the age of spaceflight and that humankind has been offered an opportunity for greatness beyond the wildest dreams of the philosophers of the past, few have yet recognized the unique importance of the planetoids in this great human adventure.

There are at least six reasons why we believe the planetoids or asteroids are uniquely important as targets for exploration and utilization:

- The asteroid-planetoids should prove to be the source of certain kinds of knowledge concerning the origin of the solar system and even of life itself—knowledge which can be found nowhere else.
- The planetoids represent a potential threat to space travelers and even to our Earth. Planetoids in the form of meteorites have hit the Earth in the past and will do so in the future. The larger, more potentially destructive ones could be purposely deflected from their orbits—by madmen seeking to "control the world"—and made to strike the Earth, destroying whole countries or even continents.
- They can be used as way stations and sources of rocket propellants for interplanetary vehicles. As we discuss in detail later, they can be mined for fuel, thus greatly reducing the cost and difficulty of long spaceflights.
- The remarkably favorable energy balance of the planetoids relative to the Earth suggests the possibility of economic transport of raw materials from the planetoids to the Earth, as well as the possibility for capturing entire planetoids and bringing them into satellite orbits around the Earth.
- They may well be the most desirable locations in the solar system for founding extraterrestrial colonies.
- They offer a possible way (and the most practical way yet proposed) for sending human colonists beyond our solar system to populate the planets of other stars.

For these reasons and many others, the United States and the world must now respond to the challenge of the planetoids.

## National and Global Goals in Space

After humankind first set foot on the moon, the inevitable question soon arose: Where will we go from there? Most experts feel that the next logical step and priority should be either the building of a manned Earth-orbiting space station or a manned expedition to Mars. However, if we look back into history for some terrestrial exploratory parallels to the forthcoming exploration of space, we can find some interesting lessons from which some worthwhile suggestions can be drawn.

If we could turn the clock back five centuries and put ourselves in the minds and shoes of the pioneering European explorers who dared to venture beyond the safe confines of the then known Earth —the Eurasian land mass—we could see a close parallel to our first steps into space. For example, we would note that the routes they chose to get to the mysterious Orient, and which ultimately took them to the New World, were not direct.

The unknown islands of the Caribbean, where Columbus first set foot in the New World, could be looked upon as having played a role similar to that of the planetoids today. A major difference in this analogy is that we have some preliminary knowledge about the existence of the planetoids, while Columbus knew little about the geography of the Western Hemisphere.

On a windy October 12, 1492, three tiny wooden ships dropped anchor off the coast of San Salvador, an island in the Bahamas. Their Italian captain, Christopher Columbus, had spotted this small island from his flagship, the *Santa Maria*, after three months of sailing westward from Spain on storm-tossed seas. He thought he had reached one of the "Indies," which had been the original goal for his epic voyage.

Although he did not establish a settlement on this island, he did so on another, larger one—but purely by accident. After the *Santa Maria* was wrecked on the shores of Haiti in a violent Caribbean storm, Columbus had no alternative but to leave forty-four of his crew members behind with their salvaged cargo, while he returned to Spain in the *Pinta* and *Nina* for more supplies. When Columbus returned to Haiti with a new cargo of horses, pigs, chickens, and vegetables, he was dismayed to find that none of the forty-four had survived. The natives' diseases and the elements had killed them.

Over a century later, when Captain Christopher Newport led his band of hardy English pioneers in three small ships to the New World, he selected an island for the first permanent settlement. The ultimate success of the venture, despite the ravages of malaria, starvation, cold, and the Indians, was due in no small part to Newport's early reconnaissance up the James River with his flagship, the *Susan Constant*, before choosing the site for the first English village, which was named Jamestown. He and Captain John Smith felt that their island would provide better protection from surprise by the Spanish and pirates on the river side with a small creek and swamp serving as a barrier to the Indians on the other side. Thus, the successful penetration of the American continent by the English came about only after they had secured a firm foothold on an island near the shores of the Chesapeake Bay.

A few years later, after Henry Hudson had claimed the discovery of the river which bears his name, the Dutch settlers who followed selected another island at the mouth of the Hudson as their foothold on this continent. Today, Manhattan Island, for which Peter Minuet paid the Indians some twenty-four dollars in trinkets and wampum, is the most valuable piece of real estate on the face of the Earth! In fact, a good argument could be made that, at present land values, Manhattan Island is worth—in dollars—more than either the entire continent of Australia or Antarctica!

Now we are about to embark on another great wave of exploration similar to that pioneered by Columbus five centuries ago. Only this time, our ships will not be sailing westward on the waters of the Atlantic Ocean but outward into the vastnesses of the cosmos.

If we can look upon the planets in our solar system as the continents of space, and the planetoids and moons as the islands, then their overwhelming importance in the coming expansion of manned and unmanned space exploration becomes more understandable. Once humankind succeeds in its first cautious landings on the moon, the next logical step may not be an expensive voyage to the "continent" of Mars to lay the groundwork for an ultimate colonization and possible exploration of that planet. Rather, it may be to land on a planetoid "island" in preparation for the more difficult landings on Mars and the other inner and outer planets.

A landing on a close-approach planetoid should be of the same order of difficulty as a landing on our moon, and a planetoid landing should be very much easier than a landing on Mars because of the relatively low gravity of the asteroid compared to that of a planet.

In this preface we have tried to provide an outer frame of reference for understanding the importance of asteroids in our solar system and what these often overlooked cosmic bodies mean to us. The next chapter covers the recent awakening of Congress, the White House, the Department of Defense, and NASA as to the threats to the Earth by stray asteroids as highlighted by the Morrison committee's Spaceguard Survey made for NASA in 1992.

## Note

1. Gunter Roth, *The System of Minor Planets* (New York: Van Nostrand, 1963).

# PART ONE

# KNOWLEDGE OF THE ASTEROIDS

# 1

# The Cosmic Shooting Gallery

During a human lifetime, there's roughly a 1-in-10,000 chance that Earth will be hit by something big enough to wipe out crops worldwide and possibly force survivors to return to the ways of Stone Age hunter-gatherers. Those are the odds of dying from anesthesia during surgery, dying in a car crash in any six-month period, or dying from cancer from breathing the automobile exhaust on the Los Angeles freeways every day. Killer asteroids and comets are out there. And someday, one will be on a collision course with Earth. Of all the species that ever crawled, walked, flew, or swam on Earth, an estimated two-thirds became extinct because of impact from space. Mankind may yet meet that fate, too. But we're the only species that can even contemplate it and, just maybe, do something to prevent it.[1]

We live in a cosmic shooting gallery. Somewhere out in the netherworld of deep space, hurling toward Earth, is a doomsday rock. The question now is not just detecting it, but what can be done to possibly nudge it off course by one means or another before it strikes the Earth and annihilates a large part—if not all—of humanity.

Such a doomsday asteroid could severely disrupt life on Earth, not only for humanity, but for the other species of plants, fish, birds, and animals. Although no astronomer has yet located the killer object (which will

29

be a mile wide or larger) headed for us, it is inevitable, according to most astronomers, that one will eventually appear. Large Earth-crossing asteroids slam into our home planet every 300,000 to a million years, which means that there is approximately one chance in 6,000 to 20,000 of a cataclysmic impact during the next half century. In other words the Earth has a much better chance of being struck by a large asteroid than most of us have of winning big in the lottery (the chances in the latter case are usually one in millions).

Dr. Tom Gehrels, a professor of lunar and planetary science at the University of Arizona who heads a team of astronomers that search the sky for such killer asteroids, says, "Eventually it will hit and be catastrophic. The largest near-Earth one we know of is 10 kilometers in diameter (or about 6.2 miles) wide. If such a thing like that hit, the explosion would be a billion times bigger than Hiroshima. That's a 'whopper!' "[2]

This new field of research in the heavens, once pooh-poohed by its detractors as laughingly paranoid, has grown in size and respectability during the decade of the 1980s. In 1989, an asteroid, a mere half-mile wide, crossed the Earth's path, coming within an uncomfortably close distance. "The Earth had been at that point (in space) only six hours earlier," a House Committee report noted. "Had it struck the Earth it would have caused a disaster unprecedented in human history. The energy released would have been equivalent to more than 1,000 one-megaton bombs."[3]

At this point in time, not knowing what else to do, but concerned for the continued well-being of voters, Congress decided to instruct NASA to conduct a study, and added into the 1990 appropriations bill the following language:

> The Committee therefore directs that NASA undertake two workshop studies. The first would define a program for dramatically increasing the detection rate of Earth-crossing asteroids; this study would address the costs, schedule, technology, and equipment for precise definition of the orbits of such bodies. The second study would define systems and technologies to alter the orbits of such asteroids or to destroy them if they should pose a danger to life on Earth.[4]

Congress "suggested" a schedule of one year for the convening of these workshops. Accordingly, NASA conducted them. The first, the "detection" group, met three times: in July, September, and November of 1991. They documented their initial work with a report issued in January 1992.[5] The second group met at Los Alamos, New Mexico (home of the atom bomb), in January 1992. They did all of their work there in three days, although the final report was not issued until February 1993.[6] We shall rely substantially upon the data assembled in these two reports, but we shall take serious exception to many of their conclusions.

The first study was conducted mainly by astronomers, who are scientists by training. Scientists are always seeking new knowledge, usually without any particular concern for how that knowledge is to be put to use. In fact, it is usually never possible to even speculate how new knowledge may be used. Certainly the person who invented the laser did not foresee its use in reproducing music from compact discs a mere generation after the principle was defined. Astronomy, in particular, has always been a passive, observant science. It has not occurred to astronomers to be concerned about such things as how to *move* the objects they study. Until the space age, such ideas were an utter fantasy anyway. Even now, moving an asteroid from its orbit is a major challenge, and the mindset of the astronomer has not been on such practical, engineering issues. This attitude plainly shows in the report of the detection committee.

The interception committee was composed mainly of people from the atomic weapons labs, with a small seasoning of "NASA types" added. Most of the weapons lab people had little or no experience in dealing with interplanetary space travel, and tended to regard incoming celestial objects as the equivalent of oversized missiles launched by an implacable Mother Nature, to be blown away by large nuclear weapons. There were some very innovative suggestions offered at the Los Alamos meeting, but there was too little time to pull all of the various ideas into a meaningful long-term plan to save the Earth.

One major criticism that can be offered for both conferences was that the mainstream scientists did not allow any room for the explosive growth of technology. They seemed to assume that the spaceships of the twenty-first century will be just like the ones they know today. In particular, they all appear to assume that it will forever be difficult and expensive to go into space. But the whole thrust of space activists is that we need to change that very assumption; it is *not* inherently difficult to get into space. It is just a technology that we have not yet conquered. The

Wright brothers, imaginative and creative as they were, probably did not envision a 747 flying nonstop from New York to Tokyo with three hundred passengers sipping wine and eating steak. And we are still just beyond the Wright Flyer stage in terms of space development today.

But is the Earth really in much danger? From what? And when?

All of these issues were the concern of the detection committee. How they answered these questions is the first place where many people disagree with the study report. Later in the book we will discuss in more detail the tale of the dinosaurs and how they met their demise. For now, we will simply say that there is no question in our minds that the theory first advanced by Dr. Luis Alvarez that they (and many other species) were all killed as a result of the impact of a giant comet or asteroid with the Earth 65 million years ago is substantially correct.[7] This theory has now been accepted by most of the scientific community, except for a rearguard action by a few paleontologists, who seem to think their professional reputations are based upon defending old ideas.

The good news is that the frequency of such spectacular events seems to be pretty rare. From the fossil record, it appears that the Earth has endured episodes similar to the death of the dinosaurs several times in 450 million years. One study claims to find a record that a major extinction event occurred about every thirty million years. The bad news is that we have no idea when our time is up. It may be in the next decade, the next century, the next millennium or the next eon. If it happens in the next decade, we are doomed, since our present technology is not yet sufficiently advanced to deal with an impact of the magnitude that annihilated the dinosaurs. But, before the end of the next century, we could advance our technology to the point that it could deal with such a threat if we decide that it is worth our effort.

There are, however, smaller catastrophes that occur much more frequently. We almost had one in 1908. At that time, something, either a comet or a stony asteroid, hit the Earth over Tunguska, Siberia. A substantial portion of its mass was destroyed in the atmosphere, but what did hit the Earth had an effect as great as an H-bomb.* Scientific estimates of

---

*When an asteroid hits the Earth's atmosphere, it is commonly said to "burn up," but this is an oversimplification. Actually, the asteroid is turned into vapor; there is no combustion involved. Although the atmosphere is heated for a few seconds, it is not usually harmed. In the case of very large impacts, however, some atmosphere (including the ozone layer) is blown away.

the destructive force have ranged from as little as four megatons to about forty megatons of explosive power. Most experts believe that the correct value is between twelve and twenty megatons, or about one thousand times the power of the atomic bombs dropped on Japan.

Trees were knocked flat over an area of several thousand square kilometers. It was later discovered that they were scorched on the side facing the epicenter, the location where the core explosion occurred. For some mysterious reason, there was no evidence of a large wildfire. One theory is that the blast wave that came along seconds after the flash simply blew the fire out! The area was uninhabited, and there are no reports of human casualties, although many deer were killed. If this event were to occur over New York, or London, or Moscow, or any major city, the devastation would be horrible.

As an aside, this was a major concern of many people during the Cold War. Such an event cannot be readily distinguished from a nuclear bomb explosion. If such a explosion had happened over either the USSR or the United States during this time of heightened tension, it could have led to the launching of a nuclear attack before the error was discovered. In fact this was one reason, among many, for the installation of a "hot line" between Moscow and Washington.

This event has been associated by some, such as Duncan Steel, an Australian astronomer, as being the impact of a piece of a comet identified as Encke.[8] The explosion over Tunguska certainly occurred during the time of meteor showers associated with Comet Encke. Others dispute this idea, claiming that a fragment of Encke would have been traveling so fast that it would have exploded at a higher altitude than what was observed. (This of course presupposes that we know what every comet is made of so that we can predict how it will act when hitting the Earth's atmosphere.) Yet other work presented at the Los Alamos interception conference leads to the conclusion that a stone only fifty yards across could easily have caused just the kind of explosion seen at Tunguska.

The reason this question is important is that the NASA detection committee study chaired by David Morrison (an astronomer at the NASA-Ames Research Center) and known by his name indicated that there are about a million asteroids fifty yards wide currently in orbits that could impact the Earth, just like the one that hit Tunguska. This does not even include the ones that may be associated with meteor streams. However, these objects are too small to be found by any search proposed by the Morrison committee. Their argument is that the probability that such an explosion will occur over a densely populated area is too small to justify looking for such objects. So they propose to look only for objects larger than half a mile wide. Such objects could disrupt an entire continent, and possibly kill a quarter of all of the people on Earth. Now *those,* the Morrison committee says, are worth looking for!

It is important to note that the project proposed by the astronomers of the search team used only *ground*-based telescopes. These telescopes would, we suppose, be controlled by the astronomers who were assembled in California to make this recommendation. We will show, later in this book, that to detect the comets and asteroids that can destroy a metropolitan area in time to deflect them will require a series of *space*-based telescopes. Whether to try to deal with such infrequent events or not is a decision that the citizens of the world need to make for themselves. It should not be made for them by a group of astronomers who wish to keep money and power in their own hands. In this book we will try to provide some basis for people to develop their own opinion on this subject.

There is yet another, smaller class of potential impacts about which we need to be concerned. These are impacts that are the size of a *small* atomic bomb, to use an apparent oxymoron. These are caused by rocks too small to penetrate the Earth's atmosphere. They explode too high in the sky to cause any significant damage. However, these impacts are detected by both the United States and Russia through our early warning systems. It is important to recognize these for what they are, or else we could make a hasty, and wrong, decision based upon the belief that an atomic attack has occurred somewhere in the world.

An example of this was cited by U.S. Air Force Colonel Simon P. Worden at the Los Alamos meeting. He said, "I want to announce that the U.S. Department of Defense sensors did detect on the first of October 1990 roughly a ten-kiloton impact. It was an airburst in the central Pacific. I note the significance of this date because had the strike occurred at that time not in the central Pacific, but in the Middle East, it could easily have been mistaken for a nuclear detonation and could have triggered very serious consequences."

There is no question that large celestial impacts are occurring on the Earth on a regular basis. But we are the first generation of the first species that has had a choice as to what to do about it. We can ignore it, as all previous generations were forced to because of their lack of space technology, or we can decide to do something to protect ourselves and our planetary home.

*What* we should do about an approaching asteroid or comet, was, of course, the main subject of a group of some ninety scientists and engineers who gathered in Los Alamos in January 1992. Unfortunately, this meeting was seriously flawed because those who attended had a poor understanding of the problem. To judge by the proceedings, very few seemed to understand the basic orbital mechanics involved. This was, after all, the first time many of the participants had been exposed to this kind of problem.

It must be pointed out that these conferences were not well run. For example, both workshops talked incessantly about Near-Earth Objects. They lumped together all celestial objects that may strike the Earth, including both comets and asteroids. There was very little effort made at either conference to differentiate between these two very different kinds of planetoids. This is comparable to an agricultural college professor telling his class that plants and animals can both be grouped together as food. Both require food, water, air, and light in order to grow. "The main difference is that plants stay in one place and animals move around." You would not think very highly of that professor, nor would you expect to learn much about animal husbandry from him. Yet this is precisely what the "experts" did at these NASA conferences on the future of Earth! We will make this difference very explicit later, and try to avoid additional confusion by using narrow scientific terms. Other examples of failed direction at the conferences will also be pointed out later.

The most essential thing in deflecting an incoming celestial missile is warning time. If we have enough notice, we can easily move mountains. It will take time and some effort, maybe a very large effort, but we can do it with an almost 100 percent chance of success. For new comets, we will have, at best, only about a year of warning, and then only if we build an advanced space-based warning system. However, given this year, we will probably be able to deflect most objects with a high probability of success. For asteroids, we may have more warning time, eventually—after the sky surveys astronomers have proposed are completed. There will still be some comets and asteroids so big, however, that they will be a major challenge to our technology for many more years.

For our purposes we have divided Near-Earth Objects or planetoids into two major classes: asteroids and comets.

## Asteroids

Asteroids exist mainly between 2 and 4 AUs (one AU equals the distance between the sun and the Earth), and have predictable orbits. We can hope to discover all of those that are dangerous in a few more years, and then expect to have years, or even centuries, to deal with the ones that are dangerous. We do not need to prepare *any* interception techniques for them *until* we find a dangerous one. There would be little reason to do so, because any defense we would prepare now would be made obsolete in a few years by the advance of technology. Carl Sagan (and others) argue that the use of large nuclear weapons, for instance, to destroy or deflect incoming asteroids would be more dangerous than the objects they are intended to intercept. Further, for most objects we can employ slight deflections, over long periods of time, without nuclear explosions, to protect ourselves.

## Comets

Comets must be divided into two sub-classes, known comets and new comets. The discussion regarding asteroids also applies to known comets, except that we must give them a wider berth, since their internal jet propulsion gives them some chance to strike more unexpectedly. New comets, on the other hand, come roaring in from great distances in the solar system, and approach us very quickly and with little warning. They are quite infrequent, but we *cannot* prepare for them in advance. If we want to take precautions, we must either have very powerful interception rockets standing by, or be willing to evacuate huge populations at short notice. There is not much point in even looking for such new comets until we have, as a global community, the political will to agree upon a standby defense, something which is, at present, lacking. We hope our readers will see the need for this.

These definitions of asteroids and comets differ slightly from the ones that the astronomers use. Some may be offended by our use of this terminology, but that does not concern us—our purpose is to make the issues clear for the general reader.

With that clarification in mind, we will return to the results of the study group that met at Los Alamos. It is important to know that the task of deflecting a celestial projectile is very difficult if the time before it will hit us is small, i.e., a month or less. If we have ten years, the task is at least one hundred times easier. Further, the means for deflection may have to be quite different for comets and asteroids, as we will discuss later. These points are readily overlooked when both comets and asteroids are lumped together.

Since the distinction between comets and asteroids was not made clear at Los Alamos, most of the participants came prepared to deal only with the case of short warning time, which led to much discussion about the use of monster hydrogen bombs to defend the Earth. Some people think that the world would be in greater danger from the inappropriate use of such bombs than we are from wayward planetoids. Our approach in this book will be to see what alternatives may exist that will be effective and also less capable of misuse.

The current guru of the asteroid-watching field is Dr. Eugene Shoemaker, a sixty-four-year-old, retired, geologist-turned-astronomer with the U.S. Geological Survey in Flagstaff, Arizona. In the 1950s he carefully studied a three-quarters-of-a-mile crater in northern Arizona which many geologists had previously believed was volcanic in origin. Shoemaker proved that the hole was created by a 150-yard-wide asteroid that slammed into the Earth 50,000 years ago. The following photograph depicts the immensity of the crater, which is approximately one kilometer across and over two hundred meters deep. Using a telescope atop Mt. Palomar, near Pasadena, California, Shoemaker has headed three U.S. teams which hunt for Earth-crossing asteroids. "They're little things and very difficult to spot," he said. "You don't see them unless you use a very large telescope, or unless they come very close to Earth. They're sort of at the threshold of detection."[9]

Congress ordered these NASA studies because the "collection of a lengthening list of Earth-crossing asteroids in recent years . . . has resulted in the accumulation of hard data," according to Dr. Clark Chapman, an astronomer at the Planetary Science Institute in Tucson, Arizona, a private nonprofit group. "The Earth is bound to be hit. Statistically, it's certain. It's unlikely that a really large asteroid will hit in our lifetime, but it's not beyond the pale."[10]

We are not alarmists. In the end, the message we want to bring to you is that the future is bright with promise. We can go forth and colonize the

asteroid belt and bring its vast riches back to serve Earth. However, we must seize the opportunities that the present affords us to expand our civilization into space or our species will indeed meet the same fate as the dinosaurs. Although the risk of death from an asteroid hit is lower than from firearms or an auto accident, it is higher than dying in an airplane crash (see Appendix E), and therefore is high enough to suggest the desirability of some sort of preventive action on the part of our government and those around the world.

By June 1991, 184 Earth-crossing asteroids were observed and mapped (at a cost to the taxpayers of less than $1 million a year). New ones are being spotted at the rate of two per month, but none is expected to hit the Earth soon. Even so, it is a statistic worth noting, because, according to a consensus of asteroid-hunting astronomers, less than 10 percent of the big asteroids have been found so far, and they are the dangerous ones that bear watching. By the time this book goes to press, over 300 Earth-crossing asteroids will have been discovered.

The most comprehensive recent search for asteroids has been conducted by Dr. Tom Gehrels, a professor of lunar and planetary science at the University of Arizona. Since the late 1980s, his twenty-six-inch telescope on Kitt Peak, Arizona, has scanned the sky with an advanced electronic detector, revealing swarms of small asteroids zipping uncomfortably close to the Earth.

On January 19, 1991, his search team spotted an asteroid about thirty feet in diameter passing between the Earth and the moon. If it had hit, the enormous velocity would have created a explosive impact equal to several bombs surpassing the one that struck Hiroshima.

## Checking Up on the Cosmic Shooting Gallery

When the NASA-sponsored team made its *Spaceguard Survey* report to the House of Representatives Science Committee in early April 1992, it laid out a plan to establish an early warning system for completing the first comprehensive census of these unique, menacing rocks in space. The down payment was $50 million, with an annual budget of $10 to $15 million per year for twenty-five years; a total cost of approximately $300 million.

The Shoemaker science team, organized by NASA, concluded that, unless some new program to search for asteroids was undertaken, only 25 percent of the estimated 2,000 threats would be found in the next decade, and that finding 90 percent of these would not occur until the middle of the twenty-first century.[11] The definition of threat, in this case, is an asteroid one kilometer (0.62 mile) or larger. Such an object would, if it hit Earth, be expected to cause *billions* of human casualties, and probably end civilization. However, if the Shoemaker committee recommendations were implemented, they believe that 80 percent to 90 percent of these objects could be discovered in the next decade.

As such a planetoid slams into the Earth at sixteen miles per second, it could explode with the force of over a million H-bombs, lofting enough pulverized rock and dust to block out the sun. The result would be a naturally produced (rather than human-generated) nuclear winter* that could wipe out humankind . . . and most of the other natural species as well.

---

*The term "nuclear winter" is used to describe the aftereffects of a thermonuclear explosion. Such an explosion would create an atmospheric dust cloud that would blot out the sun, causing temperatures to plummet and disrupting the food chain so that civilization would be wiped out.

Agriculture would be crippled and a billion or more people would die from starvation.

Outside critics, according to a *New York Times* dispatch, were quick to denounce the report as "pie-in-the-sky" busy-work project for both astronomers and weapons-makers eager to combat a new threat after the Cold War officially ended. But the primary authors of the report insisted that a wealth of new evidence compiled by biologists, astronomers, and geologists over the past decade makes such a study feasible.

The skeptics disagreed and belittled NASA with assertions that the asteroid threat is so small that NASA is "paranoid," or worse yet, conspiring to guarantee astronomers a lifetime of entitlements to keep them busy, and also provide research money to "star warriors" who wish to concoct new nukes for the cosmos. Dr. David Morrison, the cochairman of the study team from NASA's Ames Research Center in California, rebutted that criticism with the conclusion that "the risk is *real!*"

In an April 1992 editorial in the *Washington Times*, one of the skeptics envisioned a plot by a group of astronomers who offered to save the Earth from disaster with a few new powerful telescopes. That paper scorned NASA's plan as a "scam to make away with the taxpayers' money. There is no evidence that anyone in all human history has ever been killed by an asteroid!"[12]

One member of the NASA team, Dr. Richard Binzel of the Massachusetts Institute of Technology, stated that "We looked at this in the most careful, reasoned way we could. Although the threat is small, it is *not* zero. There's probably a 1–in–7,000 chance that an impact with global repercussions could happen in a person's lifetime. If you want to gamble with that level of risk, that's O.K. It's really a political decision as to whether the threat is judged big enough to warrant investment in some insurance."[13]

Dr. Morrison took the critical jabs philosophically, saying that "the asteroid threat dawned on scientists rather slowly and is difficult for the average laymen to comprehend. . . . The unambiguous fact," he asserted, "is that mankind lives in a kind of cosmic shooting gallery. This does not fall within our ordinary experience, so it takes some getting used to."[14]

The first small steps have been taken by various segments of the national bureaucracy to avert the potential asteroid threat to the Earth—which may, in time, replace the now dormant Cold War threat in its magnitude. But unlike Pearl Harbor, which shocked us into entering World War II, this time we may be prepared with early warnings of a cosmic threat from outer space.

The next chapter presents a brief overview of the first asteroid discoveries, which occurred less than two hundred years ago, and the widening knowledge of them obtained as the twentieth century dawned through introduction of larger telescopes.

## Notes

1. Melinda Beck and David Glick, "Doomsday Science," *Newsweek* (November 23, 1992): 56–57.

2. Tom Gehrels, ed. *Hazards of Comets and Asteroids* (Tucson: University of Arizona Press, 1994).

3. U.S. Congress, *House Science Committee Report*, 101st Cong., 1st sess., 1989.

4. Ibid.

5. *Report of the NASA International Near-Earth-Object Detection Workshop* (Washington, D.C.: GPO, 1992).

6. G. H. Canavan, J. C. Solen, and J. D. G. Rather, eds., *Proceedings of the Near-Earth-Object Interception Workshop* (Washington, D.C.: GPO, 1993).

7. Luis V. Alvarez et al., "Extraterrestrial Cause for the Cretaceous-Tertiary Extinction," *Science* 208 (1980): 1095–1107.

8. Duncan Steel, *Rogue Asteroids and Doomsday Comets* (New York: John Wiley & Sons, 1995).

9. Eugene Shoemaker, lecture at U.S. Science Teachers Annual Convention, Philadelphia, April 1995.

10. Cited by Shoemaker.

11. *Report of the Near-Earth-Objects Survey Working Group* (Washington, D.C.: NASA Space Explorations Division, Office of Space Sciences, 1995), pp. 1–2.

12. Editorial, *Washington Times*, 16 April 1992, p. 8.

13. Richard Binzel, AP Wire, *New York Times*, 18 April 1992, p. C1.

14. Ibid.

# 2

# The Legacy of a Roman Goddess

Between Jupiter and Mars I put a planet.
—Johann Kepler

In the middle of the eighteenth century, J. D. Titius, (1729–1796), an obscure German science professor at the University of Wittenberg, started to seek a mathematical formula that would account for the sizes of the planets in our solar system and the distances between them and the sun. After many years of observations, Titius thought he discerned a certain rhythm in the arrangement of the then known planets. His interest in finding a numerical relationship for deriving the distances to the planets led him to publish his findings in a 1772 translation of Swiss naturalist Charles Bonnet's (1720–1793) 1764 scientific work, *Observations Concerning Nature*.

Unfortunately, Titius does not always receive credit today for his discovery. Rather, the now-famous "law" of the relationship of the planets is named after another German astronomer, Johann Ehlert Bode (1747–1826). Bode had made an auspicious beginning as an astronomer in 1766 when he published a paper on the solar eclipse that occurred in that year. A few years later, Bode published a basic text on astronomy, which led to his appointment to a scientific committee whose function was to improve and refine the annual tables of the positions of the stars and the planets.

As a member of this committee, Bode later became aware of the tables which Titius had contrived in 1772 and decided to sponsor them. His arrangement with Titius resulted in what was really scientific plagiarism, for the published tables are today known as "Bode's Law." This misleading name that has come down to us 220 years later is a confusing title for the numerical relationships of the planets—since they were *not* formulated by Bode and do *not* constitute a "law" in the accepted sense; that is, they have no known physical significance and cannot be derived from more basic physical laws.

What Titius really did was to first set down the names of the planets in order of their distance from the sun. Underneath the name of each planet, Titius (Bode) placed a series of figures: first 0, then 3, then 6, with each subsequent figure doubling its predecessor. He then added 4 to each of these figures and divided the result by 10. The quotient which he obtained was the approximate distance of each planet from the sun in terms of the distance from the Earth. He used the mean distance from the Earth to the sun as a basis for determining his astronomical unit (93 million miles = 1.0 Astronomical Unit).

At the time that Bode published Titius's exercise, the only known planets were Mercury, Venus, Earth, Mars, Jupiter, and Saturn, with a questionable planetary gap located halfway between Mars and Jupiter. (Neptune and Pluto were not discovered in Titius's and Bode's lifetimes.) Here is the way the two German astronomers worked out their table, based on computations to determine planetary AUs from the sun. (It would take several years before K. F. Gauss confirmed the accuracy of Bode's Law, as is discussed shortly.)

|  | Mercury | Venus | Earth | Mars | ? | Jupiter | Saturn |
|---|---|---|---|---|---|---|---|
| Bode's Number | 0 | 3 | 6 | 12 | 24 | 28 | 96 |
| Add 4 | 4 | 7 | 11 | 16 | 28 | 52 | 100 |
| Divide by 10 | 0.4 | 0.7 | 1.1 | 1.6 | 2.8 | 5.2 | 10.0 |
| Distance in AUs | 0.36 | 0.72 | 1 | 1.52 | ? | 5.20 | 9.54 |

In 1781, less than ten years after Bode published Titius's law, Uranus was found by Sir William Herschel. It fitted into this scheme of tables with amazing accuracy, with the Bode number for the newest planet being 192. The distance in astronomical units worked out to an accurate 19.19. Confidence in the law was greatly strengthened in scientific circles.

This new confidence led many European astronomers to note that there was an obvious gap at Bode's entry of 24—with no planet to match the number. Could there be a hypothetical planet between Mars and Jupiter, as the astronomer Kepler had suggested years before and as Bode's progressive law seemed to indicate? It was puzzling, for neither Bode nor Titius originally commented on this blank in their table. According to the formula, such a planet's distance from the sun—if one existed—would be 2.8 astronomical units, or slightly more than halfway between the sun and Jupiter.

As editor of the influential *Astronomisches Jahrbuch* (Astronomical Yearbook), Bode finally became convinced that there must be a planet at a distance of 2.8 astronomical units as predicted by his colleague Titius. He was led to declare eloquently: "Is it not highly probable that a planet actually revolved in the orbit which the finger of the Almighty has drawn for it? Can we believe the Creator of the world left this space empty? Certainly not!"[1] Because Bode talked and wrote so much on the subject of the numerical tables, they finally came to bear his name exclusively. Getting his fellow astronomers to attribute the "law" to him marked Bode as the first publicity-minded scientist in modern history. In his case, the publicity assured him of a place in history which he really didn't deserve.

However, Bode did serve a useful purpose by inspiring a group of twenty-four German astronomers to organize themselves under Baron Franz Xaver von Zach (1754–1832) to make a systematic search for what they thought would be small planet between Mars and Jupiter. they divided the zodiac of the heavens into equal parts and assigned a portion to each member for study. Before they could begin their search, however, news trickled over the Alps from Italy that an outsider had discovered a small planet of the type they were seeking.

Other independent astronomers had quietly been spending many evening hours scanning the skies through their telescopes for the missing planet that would obviously be dwarfed by the giant Jupiter. One of these was Giuseppe Piazzi (1746–1826), an Italian monk and professor of astronomy and mathematics at Palermo, Sicily. On the night of January 1, 1801, while peering through his telescope, he accidentally found a strange object in the constellation Taurus. He noted that the object, which he thought to be a comet, because it was so faint, moved from night to night and, therefore, could not be a star. He followed the path of this object for six weeks until it vanished into the bright glare of the sun. He concluded that it could not be a comet, because it did not have a tail, which other comets almost invariably developed as they approached the sun due to the radiation given off by that star.

When Piazzi made his discovery, he had been in the process of preparing a new star catalog that was to become an important addition to the astronomical literature of the time. He wondered if the reason that his discovery had not sported a flaming tail could have been due to the fact that it had not penetrated close enough to the sun. The perplexed Piazzi even went so far as to write to Dr. Bode, asking him if his discovery had been seen by any of the other European astronomers. His letter to Bode and other observers also mentioned a change in the direction of the object since he first reported sighting it. But the mails were slow then (it was sent January 24 and received on March 20), and other astronomers searched in vain for this new phenomenon after receiving his letter. Apparently the new object was irretrievably lost. Yet Bode remained struck by the velocity and position of this mysterious stranger.

It fitted perfectly into the gap in his and Titius's law. Their hunch, based on a numerically ordered system, had finally paid dividends, but it was Piazzi who applied their concept and made the great discovery.

Not only was the orbit of the new object located between Mars and Jupiter at 2.8 astronomical units, but it was moving in such a way that Bode calculated it would take four years and eight months to complete a journey around the sun. He hoped that the Italian astronomer's finding would prove to be the eighth major planet in our solar system.

Bode was not disturbed by the dissertation published by the noted German philosopher, Georg W. F. Hegel (1770–1831), made just before this momentous discovery, in which he "proved" conclusively that there could not possibly be more than seven planets in our solar system. The distinguished father of the dialectical view of history poured scorn on the projected twenty-four-man search for the new planet, since he was positive it would lead to naught.

When Piazzi lost his new find in the sun's rays after his lonely vigil of forty-one consecutive nights was interrupted by illness, there was every reason to believe that it would never be recognized again. Even the greatest German mathematician of his time, twenty-four-year-old K. F. Gauss (1777–1855), wrote prophetically:

> Nowhere in the annals of astronomy do we meet with so great an opportunity, and a greater one could hardly be imagined for showing most strikingly the value of this problem (of calculating the orbits of a newly discovered object) than in this crisis and urgent necessity when all hope of discovering this planetary atom in the heavens among innumerable small stars after nearly a year, rested upon a sufficient approximation to the orbit.[2]

In the autumn of 1801, when the time came for the new body to emerge from behind the sun, its rediscovery became a difficult problem. It would have been a simple and routine matter to take Piazzi's observations made in the previous winter and run them through an electronic computer to calculate exactly where it should be on its re-entry into telescopic range even a century and a half later. But, at the start of the nineteenth century, there was no mechanical computing device that could determine the orbit of a new planet or satellite in space just month after its last sighting.

But unless some swift method of calculating the position of the so-called planet could be devised in a hurry, the mysterious object would most certainly be lost. Fortunately, Gauss came to the rescue of the frustrated Piazzi by inventing a new method based on Newton's laws—by which the orbit of a planet, planetoid, or comet could be determined (in principle) by using at least three observations of its position.

Actually, many observations spread over a period of several years were needed to obtain a precise and accurate orbit, because three observations of any newly discovered object can only result in an accurately predicted path of only a few weeks' duration. Gauss took Piazzi's observations and calculated where the little planet should be in the cosmos. By the autumn of 1801 he had figured out just where it should be and sent his calculations to interested astronomers. Gauss's calculations showed that the distance of the new planet from the sun was 2.8 astronomical units, exactly as predicted in the Bode-Titius "law."

There is a minor historical mystery as to what happened next. Who first applied the orbit calculations of Gauss to the rediscovery of the object—which turned out to be the first planetoid—and when did this rediscovery take place? Most authorities agree that the rediscovery occurred on December 31, 1801, but some give the credit to Baron von Zach and some to Heinrich W. M. Olbers (1758–1840), a German physician and amateur astronomer. A quotation from Gauss himself is of great interest: "The initial application of this method was made in October 1801, and the first clear night on which a search was made for the planet using the figures deduced from it—December 7, 1801—by von Zach brought the fugitive back under supervision."[3]

Thus it appears probable that it was von Zach rather than Olbers who rediscovered the first planetoid, but on a date earlier than that usually given: December 7 rather than December 31. The romantic notion that the rediscovery took place exactly a year after the discovery would therefore have to be abandoned.

Olbers rediscovered the planetoid on December 31, perhaps without knowledge of von Zach's observation, and his later success and fame from additional discoveries perhaps exaggerated his importance in the case of the first object. To add to the mystery, another translator of the Gauss work puts von Zach's rediscovery of the asteroid or planetoid on December 31, and Olbers's on the next evening.[4]

Piazzi, the original discoverer, named what he thought to be his "planet" Ceres, the ancient Roman goddess of agriculture and a revered deity in his native Sicily. It was determined that Ceres had a diameter of 480 miles and its mass was calculated to be about one eight-thousandth of the mass of the Earth. The dimensions are very small for a planet.

This estimate of mass was based on the measured size of Ceres plus an assumed density equal to that of the moon (3.36 as compared to water, which has a density of 1.00). The English astronomer, V. A. Firsoff, argued that if the density were actually 15 (and this is barely possible if Ceres came from the core of some planet, as astronomers believe), the mass would be almost five times greater. Firsoff makes the interesting point that the surface gravity of Ceres would then approximate that of the moon and her escape velocity would be about one-half as high as that planet's. The low average temperature of the planetoid indicated that it might possibly possess a tenuous atmosphere of heavy gases.[5] However, recent information about the asteroids makes this seem very unlikely.

But instead of the hoped-for eighth planet, the find turned out to be the first of many minor planets, or asteroids, as they are often popularly called, that started popping up all over the deep-space region between Mars and Jupiter in the form of a belt.[6]

Within the next few years, three more of these minor planets were discovered. The first of these, Pallas, was found accidentally in 1802 by Olbers, who was continuing to watch Ceres to improve its orbit description. Soon after his discovery of Pallas, he announced that both planetoids were really fragments from a single planet that had been destroyed eons ago by a giant explosion. If his theory were true, then the point of the explosion must have been somewhere in the orbital track of each of these shattered fragments. Olbers, therefore, reasoned that if other astronomers kept watch at the points in space where the orbits of Ceres and Pallas intersected, other planetoids would most probably be found.

The discovery of Juno by German astronomer C. L. Harding (1765–1834) two years later, in 1804, near the point of intersection, was a direct verification of the Olbers theory. Olbers then went to work with renewed enthusiasm, and three years later he announced that he had found Vesta. But, strangely enough, Vesta was not at the point of intersection of the orbits of the other two planetoids. Pallas and Juno were smaller than Ceres, with Juno being only 120 miles in diameter. (More recent findings have not weakened Olbers's hypothesis, however, since the disturbing gravitational effects of Jupiter and Saturn would have eliminated all traces of the meeting point [the point where the orbits intersect] unless the explosion had occurred as recently as a few thousand years ago. Olbers died in 1840 without knowing of the discovery of any new asteroid-planetoids beyond the original "big four.")

For thirty-seven years after the discovery of Vesta, astronomers believed that these four planetoids were the only ones to occupy the void at the 2.8 astronomical-unit distance from the sun. Then, in 1844, after fifteen years of searching, another planetoid, Astraea, was discovered by Karl Hencke, a German astronomer (1793–1866). Hencke's long years of patience had led to a new breakthrough in planetoid discovery at the cost of enormous labor. The preparation, every few nights, of hundreds of star charts depicting a small selected region of the sky presented a staggering task in its own right. Hencke then had to check each chart carefully to see if any "suspicious" object had moved from its previous place. If further observations revealed that the object was moving, he then knew he had discovered another minor planet.

After the discovery of Astraea, dozens of new planetoid discoveries began to fill the astronomical records, because telescopes were improving and a new photography technique was helping the sky-watchers in their search for these tiny fragments. And tiny they were by astronomical standards, for many of them were scarcely larger than a good-sized mountain on Earth.

Several of these newer minor planets were discovered within a few months by a Parisian druggist, who bought a small telescope and spotted them out of the back window of his bedroom. By 1870, more than a hundred planetoids had been found. By 1890, the discoveries had exceeded 300.

Astral photography was introduced by Max Wolf of Konigstuhl, Heidelberg, in 1891. The use of the camera attached to a telescope driven to track the stars meant that any moving body appeared as a short streak on the photographic plate while the "fixed" stars came out as small, round spots. Or, if a telescope was "locked" on a planetoid, the stars would all appear as streaks, while the planetoid was distinguished by a round image. This latter method provided an advantage to the astronomer who spotted a faint object, since it could then build up a stronger image on the plate than if it were followed as a trail.

When this method was applied in earnest to the search for new planetoids, the resulting flood of discoveries created a vast confusion of new names and claims. Obviously, some sort of international monitoring agency was needed immediately to bring order out of the chaos of discoveries. So, the supervision of the results of the great planetoid hunt was placed in the hands of the *Recheninstitut* of Berlin, the premier scientific institution at that time, which promptly laid down certain rules for further discoveries.

Any new planetoid finder was granted the privilege of naming his discovery, which was recorded in the German register, *Kleine Planeten* (Small Planets). The new find was given a temporary designation until its authenticity was officially determined. An orbit of a new planetoid was verified after at least three perihelion passages (revolutions around the sun) had been completed, and it was determined that the discovery was really a new object and not just an old one making its regular rounds. Then it was assigned a new number in the order of its discovery. The number followed the name assigned to it by its original discoverer—for example, Eros (433).

This chapter has provided a brief history of the early discovery of asteroids at the beginning of the nineteenth century and the refining of discovery techniques as more sophisticated telescopes were introduced in the twentieth century.

The next chapter covers the quixotic and sometimes humorous, even ludicrous, names given to each new asteroid discovery by its founders. Unlike the nine planets in our solar system, the discovery of thousands of asteroids strained scientists' abilities to bring some semblance of order to the naming process.

## Notes

1. Cited in Donald Cox and Dandridge Cole, *Island in Space: The Challenge of the Planetoids* (Radnor, Penn.: Chilton, 1964).

2. K. F. Gauss, *Theory of the Motion of the Heavenly Bodies, etc.* (Boston: Little, Brown & Co., 1857), p. ix.

3. Ibid.

4. H. H. Turner, *Astronomical Discovery* (Berkeley: University of California Press, 1963).

5. V. A. Firsoff, *Our Neighbor Worlds* (New York: Philosophical Library, 1953).

6. The official international designation is "minor planet." Although "asteroid" (little star) was suggested by the great astronomer William Herschel, and is more commonly used by English-speaking astronomers, many astronomers would prefer the more logical scientific term of "planetoid" to describe these cosmic objects.

# 3

# The Explosion of New Discoveries

Before the International Astronomical Union (IAU) got into the act (of naming new finds in space) there was chaos.
—Leif Robinson, editor of *Sky & Telescope Magazine*, 1994

The naming of the new planetoids soon became a growing problem. Originally, they were all given feminine names taken from classic mythology, but these were soon exhausted. Then, leading figures from Wagnerian operas and Shakespeare's plays were selected either by the discoverer or sometimes by the person who computed the orbit. When these theatrical names were also used up, as the number of new planetoids continued to increase, the astronomers began naming them after their wives, sweethearts, fellow astronomers, friends—even their pet dogs and cats.

By custom, most of the planetoids have been given Latin feminine names (if they lie in the region between Mars and Jupiter). Since the astronomers have long since run out of the conventional supply of well-known classical feminine names, they started adding the Latin feminine ending, -a, to masculine names to label their newest planetoid finds. The results have often been humorous and tongue-twisting concoctions of well known and unknown names, such as Rockefellia (904), Carnegia (671), Limburgia (1383), Arnica (1100), and Geisha (1047).

51

Even three of the astronomers associated with the finding of the first planetoid, Ceres, have been commemorated by naming minor planets 1000, 1001, and 1002 after them. They are called Piazzia, Gaussia, and Olbersia.

American astronomer E. E. Barnard (1857–1923), who discovered the fifth satellite of Jupiter, is also represented by Barnardiana (819) and his wife, Rhoda, also got into the act with No. 907 being named after her. Johann Palisa (1848–1925) of Vienna, who was a noted turn-of-the-century planetoid hunter, was conspicuously honored by his fellow astronomers. Three new planetoids were named after his outstanding personal characteristics: Probitas (902), Perservantia (975) and Hilaritas (996).

Although planetoids have been named for Greek heroes, places, astronomers, and astronomers' wives, daughters, dogs, and cats, the pioneers of rocketry and astronautics have not generally received this honor. Only one planetoid has been named for a spaceflight pioneer, and significantly, this is No. 1590—Tsiolkovskaja, named for the Russian father of cosmonautics, Konstantin Eduardovich Tsiolkovsky (1857–1935). There is no planetoid named after the Hungarian, Oberth, and none named after the American, Robert Goddard, the other two giants in the triumvirate of astronautical greats. (Once again we see confirmation of an earlier recognition of the importance of spaceflight among Russian scientists.)

In the late nineteenth and early twentieth centuries, two new special groups of minor planets were discovered distributed in Jupiter's orbit equidistant from the sun and Jupiter. At this point, the astronomers decided to abandon the traditional Latin feminine names for these finds. They were simply running out of the available supply of classical heroines. Since the reserve of classical male names was almost untapped, the astronomers took the *Iliad* off the shelves and named planetoids in both of these groups after heroes of the Trojan War. It would have been ideal if one group of these latest finds had been named after Greek warriors and the other for Trojans, but some early mistakes spoiled this possibility.

In 1908 when No. 588 was discovered in Jupiter's orbit sixty degrees ahead of the giant planet, it was named Achilles. When No. 617 was discovered in the same year, some sixty degrees behind Jupiter, forming a second equilateral triangle with our largest planet and the sun, it was natural to name it after the friend of Achilles and it became Patroclus. Then came No. 624, found orbiting near Achilles, and named Hector for the

Trojan hero. From then on, an effort was made to keep the Greeks in one camp and the Trojans in the other, but the fifth columnists had already secured their positions.*

The most distant wanderer found to date is Hidalgo (944), discovered by German astronomer Dr. Walter Baade (1893–1960) in 1920. It comes in to a distance of two astronomical units at perihelion (close approach to the sun) but goes out to 9.5 astronomical units at aphelion (furthest distance from the sun). This is almost ten times the Earth's distance from the sun and goes well beyond the orbit of Jupiter to the orbit of Saturn.

After the close of World War I, several Russian astronomers made a friendly gesture to the United States unique in the history of naming planetoids. It dramatized the universality of international scientific cooperation. During the Russian famine of 1919, word trickled back by mail to some American astronomers that several of their Russian colleagues were near death from starvation. News of their plight was relayed to the future Quaker president, Herbert Hoover, who was in charge of the American Relief Administration (ARA), and he promptly saw to it that food, clothing, and supplies—donated by the American astronomers—were dispatched to professors in Moscow and the Crimea.

Because these Soviet astronomers felt that their lives were saved by the generosity of their American colleagues, as a token of their gratitude, they named a newly discovered minor planet ARA after the organization that kept them from starving. This would be one of the few outstanding examples of international, altruistic understanding to take place in the ensuing forty years between scientists of the Soviet Union and the United States. (More recently, in a similar friendly gesture, Russian astronomers named one of the prominent geographical features discovered by *Lunik III* on the moon after the American inventor Thomas Edison.)

As a whole crop of new asteroids kept popping up in the latter half of the twentieth century, the keepers of the asteroid names ran out of god-

---

*Hector was soon followed by Nestor (659) and Agamemnon (911), who quite properly were placed with Achilles. Later, Ajax (1404), Odysseus (1143), Menelaus (1647), Diomedes (1437), and Antilochus (1583) joined the Greeks, and Priamus (884), Anchises (1173), Troilus (1208), and Aeneas (1172) joined the Trojans. All of these are fairly large and there are undoubtedly many more of moderate and small size which have not yet been discovered in the Trojan groups. Achilles is about 150 miles in diameter—not large for a planet or a continent, but a very respectable size for a space island.

desses and other appropriate names for these minor planets. Soon there was a Soviet-discovered asteroid named after Vladimir Lenin (Wladilena) and one after President Herbert Hoover (Hooveria). All four of the Beatles each had an asteroid named after them and the Finnish Olympic runner, Paavo Nurmi, had his name put on one, as did assorted Japanese lakes and rivers. Even such musicians as Mozart, Debussy, and Leonard Bernstein have been honored in the skies.

The people discovering asteroids soon started fighting among themselves about proper names for their finds, but they also found themselves fighting with the "moon people," who had begun taking the best asteroid names and giving them to newly discovered moons found by the space probes *Voyager I* and *Voyager II*.

One of these interplanetary turf battles flared in 1989 when *Voyager II* found six small moons circling Neptune, two of which were named Galatea and Larissa; names which already belonged to two asteroids. The controversy was officially settled in early October 1991, when a compromise was reached allowing the moon people to keep names but promise to not do it again. The asteroid people, however, groused that with no new space missions planned to the outer planets in the near future, it was unlikely that any new moons would be discovered anytime soon in our solar system. "We wonder if we've been had," said one frustrated asteroid aficionado.

This hassle marked the latest strife to befall the International Astronomers Union (IAU) which was founded in 1919, just after World War I, to encourage international cooperation among the world's astronomers. Fortunately, the IAU—based in Paris—had become recognized as the organization with the last say on the naming of new celestial bodies discovered by Earthlings.

The IAU has forty separate commissions naming all objects that fall under their respective purview. For example, Commission 27 handles the galaxies and the stars, while Commission 5, covering the asteroids, has separate committees to name comets and asteroids and to determine when a moon is in fact a moon with a verifiable orbit. This committee, however, doesn't get the privilege of actually naming a moon. That job falls to the IAU's Working Group on Planetary System Nomenclature (WGPSN)—the moon people.

Although this continuing bureaucratic struggle over naming new finds continues, Leif Robinson, the editor of *Sky & Telescope Magazine*, for whom an asteroid has been named, pointed out that "before the IAU, there was chaos."[1]*

By late 1991 there were 4,960 known asteroids in the solar system and a predicted doubling of that number by the year 2000. Because every month the IAU's asteroid committee receives 200 to 300 new finds for naming, there has been a fear that it will soon run out of names. "We worry about that," says Brian Marsden, an asteroid committee member and director of the Minor Planet Center located at the Harvard/Smithsonian Center for Astrophysics in Cambridge, Massachusetts. "The suggestion was made that we stop naming [the asteroids] after 5,000. Just give them a number, but our members were overwhelmingly opposed. If we don't do it through the traditional channels, someone else will do it anyway." (He was referring to the "star registries" to which certain ego-conscious persons pay a fee to have stars named after them.)

Today, the asteroid discoverers do follow a certain set rules; for example, assigning names that are pronounceable do and not have more than sixteen characters, preferably in one word, as well as words that are not deliberately obscene or in bad taste. It has also been agreed that to name an asteroid after a political or military hero or event, the naming should not occur until one hundred years after the person has died or the event has taken place. This last rule went into effect when the Soviet astronomers in the Crimea began naming their asteroid finds after famous Soviet generals, like Georgi Zhukov. This process became objectionable in international circles when the Russians wrote up their heroes as conquerors of the "nasty German armies." Other national astronomers felt that these names did not serve any astronomical purpose, so to promote future cooperation and friendship among global astronomers, this type of naming is frowned upon. Also, once named, an asteroid moniker cannot be rescinded. For example, when an Argentine astronomer named a planetoid after that country's former first lady, Eva "Evita" Peron, and then

---

*For instance, when the twelfth asteroid was discovered in 1850 and it was named Victoria, the Americans objected, because they felt she was a "queen" and not a "goddess." American astronomers had opted for the name Cleo, and the controversy prevented the *Astronomical Journal* from accepting the former name for seven years. Arguments zipped back and forth across the Atlantic during those years until "finally, the Americans gave up," Marsden said. The name Victoria stayed.

tried to rescind it, the other world astronomers objected, so No. 1569 remained "Evita."*

## How Many Are Out There?

Today more than 700 planetoids are known to be in orbit in our solar system, with over 1,700 being confirmed and officially listed in the astronomical tables. The rest are struggling along with their temporary numbers, waiting to be admitted to the coveted inner circle.

It has been reliably estimated that there are at least 47,000 unknown planetoids bright enough to be photographed with a 100-inch telescope when they approach nearest the Earth. In fact, an astronomer who has access to one of these big telescopes often has difficulty not finding a planetoid when he sits down in front of the eyepiece of his instrument. Russian astronomer S. V. Orlov has estimated that there is a grand total of at least 250,000 planetoids in our solar system with a diameter of 3,000 feet (one kilometer) or larger! (This estimate could be too high by a factor of four or five, but the number is still large.)

When American astronomer Seth Nicholson (1891–1963) was searching for new satellites of Jupiter with a 100-inch telescope, he picked up a remarkable total of thirty-two planetoids on his photographic plates. They were nothing more than nuisances to him, because each one had to be painstakingly eliminated as a possible moon of Jupiter. A planetoid may parallel Jupiter's orbit for a few days like the rest of that planet's satellites, but after a week or so, its independent motion will reveal itself and it can be labeled as another minor planet.

The first planetoid to achieve fame through its close approach to the Earth was Eros (433), which was discovered in 1898 by German astronomer Gunther Witt of Berlin. Eros has an elongated orbit and comes much closer to the sun than Mars at one end of its sweep around our solar

---

*One controversial name belongs to "Mr. Spock," a little tabby cat named after the "Star Trek" TV-program and movie's first officer with the pointed ears. When an American astronomer spotted a piece of a hitherto unnamed rock zooming between Mars and Jupiter, he wanted his pet cat to receive eternal recognition. So he named it "Mr. Spock." There have been more objections to that asteroid name than any other so far, but with valid names running out, we can look for more "Mr. Spocks" in the future.

system. Sometimes it sweeps in as close to Earth as 13.8 million miles. For more than thirty years, this four-mile-wide, fifteen-mile-long, football-shaped planetoid, tumbling end-over-end, reigned supreme as our nearest neighbor. In 1931, it swished in to within 17 million miles of the Earth—much closer than any planet.

Eros was the first asteroid found to move not in an orbit consistent with and beyond the orbit of Mars, but in an orbit which crosses that of Mars. (It had become the rule to bestow female names on the newfound asteroids, but when it was discovered that Eros crossed the orbit of a major planet, it was given a male name. Following that, about two dozen orbit crossers of Mars were discovered and given male names. These are often referred to as "male asteroids.") For some time these asteroids were thought to be the exception to the rule. However, recent discoveries show that there are at least hundreds and probably thousands of such orbit-crossing bodies.

The discovery of Eros enabled astronomers to check their measurements of astronomical quantities, and compute a more accurate value for the mean distance to the sun. This vital cosmic yardstick is important for assessing distances to the planets and other objects far out in deep space.

The crossing of planetary orbits by these asteroids has been likened to a bridge crossing a highway rather than to two highways intersecting. This is because many asteroid orbits are on angles inclined to Earth's orbit. It is possible for an asteroid to be crossing our path while remaining millions of miles above or below the Earth. This reduces the danger of an Earth collision from these bodies, but there are now so many Earth crossing bodies that the danger is still significant.

After the discovery of Eros in 1898, the Earth had no more close calls (at least none that were known) until 1911, when Johann Palisa (1848–1925) of Vienna discovered a small object of less than three miles in diameter whose orbit came within 20 million miles of the Earth. This was Albert (719), which has a perihelion of 1.19 AU (astronomical units) and an aphelion of 3.98 AU. Two more objects with similar orbits were soon found—Alinda (881) and Ganymede (1036). Both Albert and Alinda were lost for some time, but have been rediscovered.

The discovery of these small planetoids moving in elongated orbits that come far inside the main zone between Mars and Jupiter has led to some interesting speculations about the origin and stability of our solar system. These tiny objects can only be spotted when they are very close to us. Because they shoot by the Earth so quickly, not enough accurate observations have been obtained to determine the precise orbits of some.

Some of the unanswered questions which these newest finds have encouraged our astronomers to ask include the following: Why are their orbits more like the comets' than the other planetoids? Why do their paths differ so radically from the larger planetoids located in the conventional planetoid belt? Are they really former comets that have left their tails behind to become minor planets? Did a close encounter with the powerful gravitational pull of Jupiter distort the orbits of these formerly "normal" planetoids into skewed abnormal paths around the sun? Only close examination of these small chunks of cosmic debris will provide us with the answers to these perplexing questions.

When a planetoid is scheduled to be in a certain position in the sky as seen from one point on the Earth's surface, it will appear to be displaced from that position when it is viewed by an observer from a different spot on the Earth. This apparent displacement will pinpoint the distance to a close-approach planetoid more accurately than we could measure the distance to the sun or the planets. Then, with the aid of this measurement and Kepler's Third Law (which relates orbital time periods to distances from the sun), we can calculate the distances from the sun of other planets.

## Eros and Icarus Do a Dance in the Heavens

In 1975 Eros came within 14 million miles of the Earth, making it a much easier target than Mars for future manned astronaut explorations beyond the moon. Since Eros made its first detected close pass by the Earth, others have been seen to come even closer. In March 1932, Amor came within 10 million miles of us. This record did not last long, however, because Apollo cut this distance nearly in half when it whizzed within 6.5 million miles of our planet, with part of its orbit falling slightly inside our own. In 1936, Apollo's record was broken by Adonis, which swept by us at the uncomfortably close distance of 1.3 million miles, only five times the moon's distance from the Earth.

Until very recently, the record for the closest approach of any celestial body to our planet—except for our own moon and the meteorites—was held by an asteroid known only as 1989 FC. On March 23, 1989, this body passed within 450,000 miles of the Earth, which is roughly twice the distance to the moon. It was estimated to be the size of a giant aircraft carrier, traveling at the speed of 42,000 miles per hour. If it had hit the Earth, it would have made a crater several miles in diameter, or perhaps caused a tidal wave three hundred feet high if it landed in the ocean. Such a wave would utterly devastate any coastal area it hit. This asteroid was not discovered by earthbound astronomers until three weeks after it had passed our planet. If a half-million-mile miss does not impress one as being a close call, just think of the consequences in terms of time: it missed us by only six hours! The outcome if it had hit us would be nearly unimaginable.

Even though 1989 FC no longer the record for close approach, it will remain in memory, since its discovery is the event that first precipitated Congress to pay attention to the asteroid menace. Congress encouraged NASA to begin to study the problem, creating a positive result from this asteroid threat.

Forty years before the discovery of 1989 FC, a previous asteroid find should have provided an early warning of things to come, but it did not. On June 26, 1949, this vanguard planetoid appeared on a photographic plate taken by astronomer Walter Baade with the forty-eight-inch Schmidt telescope on Mt. Palomar in California. The trail on the plate was so long that in eleven hours it would have covered an apparent distance equal to the angular diameter of the moon. The trail was like an arrow without a point, showing a line of motion, but Baade was unable to determine the asteroid's direction. He was concerned that this latest find would be lost.

Several fellow astronomers helped to compute the orbit of the newly found object and make further observations of its path. The irregular orbit was determined to be so eccentric that at its closest point to the sun it passed 17 million miles inside the orbit of Mercury. At aphelion, it swept some 42 million miles beyond the orbit of Mars.

Baade dubbed his find, the only known planetoid that passes inside the orbit of Mercury, Icarus, after the rash and impulsive aviator of Greek mythology.* In 1950 and 1952, Icarus was observed again as it passed near the Earth, and there is no danger that it will ever be lost, since its orbit has been plotted with great accuracy. The asteroid passed within

---

*Icarus and his father Daedalus, according to mythology, tried to escape from Crete by means of wings of feathers attached to their bodies with wax, but Icarus flew so high that the wax melted from the heat of the sun, and he fell to this death in the Aegean Sea.

four million miles of Earth in 1968, the closest approach of that decade for any planetoid of known orbit.

Astronomers have estimated that when Icarus is closest to the sun, its surface glows with a dull red color, while six months later, in the extremely cold void between Mars and Jupiter, it is pitch black. Because it comes within some 8 million miles of Mercury, this planetoid, properly equipped with measuring instruments, could eventually help us obtain a better value of the mass of that planet, the closest planet to the sun. Icarus may also help us confirm Einstein's theory of relativity more accurately than it has been to date by providing measurements of the motion of the perihelion of Mercury's orbit. One of the obscure consequences of Einstein's theory is that the orbit of Mercury should change more than predicted by Newton's laws of motion. The greater change has been observed, and Icarus offers verification of Einstein's work.

## Three Generations of Asteroid Discovery

The world is now into its third generation of asteroid discoveries. The first generation began with the invention of the telescope, which enabled asteroids to be seen for the first time, and humankind began to catalog this strange new sight in the heavens. As noted earlier, photography was introduced in 1891. The use of time exposure permitted asteroids too faint to be seen by the human eye to be recorded on film, and their orbits determined, in a second generation of discovery.

The third generation of asteroid discovery began with the use of Charge Coupled Detectors (CCDs), an electronic equivalent to film, in astronomy. This generation is now several years old. (These are the same devices used in home videocameras. The ones the astronomers use are even more sensitive and sophisticated than the ones in a camcorder.) These detectors can be even more sensitive than photographic film, which has improved dramatically since it was first introduced a century ago. In fact, the present generation of CCDs is approaching the theoretical limit for detecting light. This means that we cannot hope to improve our search techniques without better detectors. The only avenues open to improve our search is to use much larger telescopes, or to get the telescopes into space. Or, best of all, to do both.

The use of CCDs has another advantage not readily apparent: The gathered data are directly computer compatible, i.e., a computer can "look through" the telescope. This means that the process of finding new images in a crowded star field can be automated, reducing the drudgery that the astronomers have to endure. The use of these new techniques has dramatically increased the rate of finding new Earth-approaching bodies. The few search teams are now finding several such asteroids per month, a vast improvement over even a few years ago. However, even more improvement is needed if we are to find all of the suspected threats within a decade or so in order to give us more lead time for perfecting methods of deflecting those coming close to Earth.

This chapter offered a cursory overview of the fascinating but helter-skelter naming of newfound asteroids, plus the attempts to bring order out of that chaos. The most important points, however focus on some recent attempts to speed up new discoveries of these elusive objects.

In the next chapter, we offer a broad comparison of the similarities and differences between the predominantly icy comets and the rocky asteroids and how our knowledge of their peculiar eccentricities can aid us in the better understanding of each type of cosmic object and how they relate to one another.

## Note

1. Leif Robinson, Editorial, *Sky & Telescope* (April 1994): 12.

# 4

# Cosmic Cousins

The principal thing we will want to bring back from space exploration
is more knowledge.

—Dr. Edward Teller, "Father of the H-Bomb"

Astronomers and others who have been recruited into the group discussing the defense of planet Earth from celestial missiles have fallen into
the habit of calling these threats Near-Earth Objects, (NEOs) because,
unlike the planets or stars, they come close enough to actually impact
Earth and cause a catastrophe. However, we think that this term is quite
misleading, since these objects are *not* at all close to Earth, at least until
minutes before impact. Most of the time they are millions of miles away,
vastly farther away than the moon, for example. Thus, we propose the
acronym "COTE," Celestial Object Threatening Earth, as a more apt
description for these bodies.* We will use this term interchangeably with
the astronomers' term NEO, although we prefer our own term.

---

*A typical COTE will approach the Earth at something like 35,000 to 40,000 miles
per hour. They will be farther away from Earth than the moon is until several hours before
they collide with Earth.

# Where Did the Asteroids Come From?

In an article in the October 1992 *Scientific American*, Dr. Richard Binzel, an asteroid specialist at the Massachusetts Institute of Technology, stated that he and his colleagues were now convinced that the asteroids are remnants of a planet that failed to form primarily because of the chaotic effects of the gravity of the nearby giant Jupiter. The influence of this giant planet prevented the gathering of these planet fragments which condensed from the original cloud of gases into one place as happened with all of the other planets. Instead, a number of much smaller bodies formed in the place that Bode's law (discussed in chapter 2) predicts for a planet. How many smaller bodies were formed is still a source of dispute.

The oldest theory for the source of the asteroids was originally proposed by H. W. M. Olbers, who, as was mentioned, discovered the second planetoid, Pallas, in 1802, and the fourth, Vesta, in 1807. Olbers suggested that these four objects must be fragments from a larger planetoid that had somehow been destroyed. Olbers's planet explosion theory resulted naturally from the prediction by Bode that a planet would be found between Mars and Jupiter, and the subsequent discovery of four planetoids rather than the single predicted object. Based on the percentages of iron and stony meteorites which have been observed to fall, Zavaritskii, a Russian astronomer, estimated that the parent planet would have had a hard outer crust of about 1.5 percent of the total planetary radius.

R. A. Daly, an asteroid scientist, estimated that this protoplanet (i.e., minor planet), would have had a mass of about one-fifteenth that of the Earth, with a radius of three thousand kilometers (1,850 miles) and an iron core radius of one thousand kilometers (620 miles).

The most plausible explanation for why this protoplanet split up is that it broke one of the rigid laws of the universe: On one of its regular orbits around the sun, it found itself venturing too close to colossal Jupiter. The gravitational pull of our solar system's largest planet (or possibly a collision with one of Jupiter's many moons) then shattered it into thousands of smaller pieces. This body, with a mass smaller than its neighbors, had crossed into the fatal planetary "danger zone." Any orbiting object that moves into such "no-man's land" in space is breaking the law known as "Roche's Limit," which describes the closest distance at which a natural satellite can orbit its primary.

Gerald P. Kuiper, a Cornell University astronomer, once proposed a multiple protoplanet theory in which perhaps five to ten medium-sized bodies formed in the planetoid belt between Mars and Jupiter. He argued that the gravitational pull of Jupiter prevented the formation of a single large planet in the region of the planetoid belt. Subsequent collisions between two or more of these mini-planets produced the large number of existing planetoids and the thousands of other bodies which have struck Earth's moon and the other planets in past ages. Collision and fragmentation still account for some of the planetoids and meteorites, according to this theory, but many were the result of a failed "building up" process rather than a successful "breaking down."

Fragmentation of the original group of asteroids has resulted in "families" of asteroids. This was pointed out in *Islands in Space*, and is still the subject of scientific study. The idea has been that, by calculating the orbits of individual asteroids, and calculating backward in time, it would be possible to find their common point of origin. In 1917 Japanese professor Kiyotsuygu Hirayama took on the task of of analyzing the orbits of the 790 planetoids which had been cataloged up to that time. He started with an exciting prospect; if he could show that all of these asteroids had originated in one area of space, the protoplanet explosion theory would be confirmed. This was a formidable challenge, since the project was undertaken long before the advent of the modern computer. Hirayama did not complete his study until 1928, by which time another 256 asteroids had been added to the list. Unfortunately for his exciting prospects, "Detective" Hirayama did not find one unstable planetary culprit at the end of his long trail of calculations, instead he found what appeared to be *five* planets.

Hirayama found that many of the known planetoids could be grouped into five families according to their orbit characteristics:

1. The Flora family at 2.2 AU (57 members);

2. The Maria family at 2.5 AU (13 members);

3. The Koronis family at 2.9 AU (15 members);

4. The Eos family at 3.0 AU (23 members); and

5. The Themis family at 3.1 AU (25 members).

Of course, this only accounts for a fraction of the total number of asteroids, so their origin at a few points in our solar system was not established.

Hirayama's work is being continued by people such as Jim Williams of the Jet Propulsion Laboratory,[1] who has identified 104 mostly small planetoid families. This is probably the most extensive list. Vincenzo Zappla and his co-workers in Italy have another list of twenty-one families that they consider to be statistically reliable. The three largest families are among those identified by Hirayama in his pioneering study. There are more than 5,000 currently identified asteroids, and some 860 are considered by scientists to be in families. Of these, 600 belong to the three large families, Koronis, Eos, and Themis. The two planetoids photographed by the spaceship *Galileo* on its voyage to Jupiter belong to families. Gaspra is associated with Flora, and Ida belongs to the Koronis family.

The answers to some of the major problems of cosmology and cosmogony—the nature and origin of the universe—may be found on the planetoids. Since Mars, Venus, and most of the planets in our solar system have atmospheres similar to that of the Earth, their surfaces have undoubtedly been transformed in the past by eons of weathering forces—winds and possibly oceans. Even the moon has suffered from erosive weathering caused by extremes in temperature and solar radiation, and by a vast number of impacts from comets and meteorites of all sizes, from mountains to dust particles. Also, the moon is covered with a thick layer of regolith (dust) which obscures its ancient surface from spectrographic observation.

The relatively tiny, faster-spinning planetoids, however, should have less violent temperature changes and suffer less erosion from thermal stresses. Because so many of them spend most of their lifetime at a greater distance from the sun than does the moon, they probably experience less severe showers of solar radiation.

Planetoids have enhanced the development of the science of celestial mechanics by aiding in the solution of problems associated with their motion and by improvements in computing the astronomical unit (through careful observations of the close-approach planetoid, Eros). From a scientific point of view, the planetoids also are a key factor in determining the mean density of matter in the solar system. This bit of information is extremely important in establishing how mass is distributed throughout the solar system, which is vital to an understanding of the evolution of the universe. The number, size, density, composition, and distribution of the minor planets are important missing links in the chain of scientific knowledge about the cosmos.

## Asteroids and Meteorites

In late May 1992, two cosmic chemists from the University of Arkansas reported that a spray of meteorites struck the Earth about 300,000 years ago after an asteroid broke up. The conclusion of Drs. Paul H. Benoit and Derek W. G. Sears, based on evidence collected in Antarctica and reported in the journal *Science*, reopened one of the more contentious issues in planetary science: whether or not meteorites found in the icy Antarctic are different from those found elsewhere on the planet.

If such meteorites are indeed different, this peculiar group found in the South Polar region may have something important to tell us about the solar system's environment and how it has changed over the passing millennia. The two scientists believed that a mother asteroid broke up eight million years earlier and that it took over seven and a half million years for the meteorite debris to shower upon the Earth. They feel that the bitter Antarctic cold and ice makes that region an ideal collecting trap for meteorites and that underground rivers on that continent could carry a space rock for miles. When the rock hit an obstacle, it would be thrust upwards to where surface discoveries have been made in recent times.

These Antarctic meteorite specimens were found near the Allan Hills and the Yamato Mountains, where the long-buried space rocks had been swept up to the surface by icy streams. After the ice is worn away by wind and weather, the embedded meteorites that had been entombed for millions of years are exposed to the geological collectors.

More than 10,000 such meteorites have already been gathered in the Antarctic to date. Dr. Benoit reported that "During last season alone (1991), some 600 meteorites were harvested there. Meteorites give us a powerful tool for interpreting solar system history, and there seems to be plenty more where we've harvested thousands already."[2]

## The Meteorite-Asteroid Relationship

Most meteorites are chunks of asteroids, large and small, that have broken off the primary mother body. Most of them burn up as they enter the atmosphere of the Earth or the planets Mars and Jupiter. Only a few manage to survive the hot temperatures of re-entry and make a large crater on land or a big splash in the ocean. The Earth has been hit by many more meteorites than asteroids in its history, simply because there are many more of the former. But even though they are smaller, they can do major damage if and when they strike.

Benoit and Sears are fairly certain that one type of unusual meteorite which they harvested came from a single asteroid. The stony meteorites on which they focused their attention came from a class known as H5 chondrite. Although the H5 type are commonly found all over the Earth, those analyzed in the Antarctic displayed a feature not found anywhere else on the planet: Those meteorites must be heated with energy from laboratory radiation to extraordinarily high temperatures before they will emit a thermoluminescent light flash. While ordinary H5 chondrites glow brightly for a few seconds after being heated to 190 degrees Celsius (374 degrees Fahrenheit), some of the Antarctic H5 specimens only glowed when heated well above that temperature. (They deduced that the reason for this phenomena was the difference in mineral composition of the two groups.) They believed that while the meteorites were whirling about in the cosmos for eight million years after splitting off from the parent asteroid, they had been bombarded by a heavy dose of cosmic rays, producing the radioactive element chlorine-36 in their outer layers. After hitting the Earth, they were shielded from more cosmic rays, which allowed the scientists to test the remaining chlorine isotope to estimate the date they splashed down on our home planet.

Significantly, among the recent meteorites found in the Antarctic during the past ten years, scientists have found fragments of our own moon which were blasted loose by meteorite impacts on that satellite, and then hurled earthward. Although the identification of these rocks as lunar in origin was controversial at first in astrogeological circles, direct comparison of their mineral structure with that of the rocks brought back to Earth by our astronauts eventually laid those doubts to rest.

Some scientists believe that another group of Antarctic meteorites reached our planet from Mars. On August 7, 1996, Daniel Goldin, NASA

Administrator, declared at a Washington news conference that the scientific evidence gleaned from a meteorite found twelve years before in the Antarctic showed "an infestation of germ-shaped structures. . . . It is a day to go down in history."[3]

President Bill Clinton announced simultaneously that he would be convening a summit in the nation's capital during November 1996 to "discuss how America should pursue answers" to the question raised by the possibility that simple bacteria could have existed on Mars three and a half to four billion years ago.

The four-and-a-half-pound historic find, labeled Allan Hills 84001 (after the spot in the Antarctic near Mt. Nansen where it was discovered in 1984), was probably kicked up by a violent asteroid splashdown on Mars 17 million years ago. After circling the solar system for eons, it landed on the Earth 13,000 years ago. It took until 1994 before an analysis of the meteorite's content was complete. The results showed signs of what could be microcarbon life which existed on Mars some 3.5 billion years ago, when the planet was warmer and wetter.

To emphasize the importance of this discovery that we Earthlings are not alone in our solar system, President Clinton commented that, "It [the meteorite] speaks of the possibility of life. . . . If this discovery is confirmed it will surely be one of the most stunning insights into our universe that science has ever uncovered."[4]

## The Differences between Comets and Asteroids

Now that we have discussed the origins of asteroids, we need to say something about comets and why they present such a very different threat to the survival of life on Earth. Like the asteroids, comets are also believed to have been left over from the very earliest beginnings of the solar system.

In April 1976, Dr. Thomas C. Van Flandern of the United States Naval Observatory issued the results of the study of sixty comets that indicated they originated from the explosion of a giant planet that once existed between Mars and Jupiter five to seven million years ago. His findings, drawn from a year of calculations of the comets spread over a zone extending about 100 million miles wide, were in line with a similar theory proposed in 1972 by M. W. Overden, an astronomer at the University of British Columbia.

Van Flandern's computer plotting of the comet orbits showed they do not intersect exactly at one spot in the asteroid belt of our solar system because of gravitational influences of the galaxy. This finding seemed to confirm the long-held theory that the once missing giant planet existed up until six million years ago. "At the time," Van Flandern said, "the planet exploded—accounting for most, if not all comets, the asteroid belt, and many meteorites." Although he failed to speculate on what caused the explosion, he did postulate that the comets, asteroids, and most meteorites were born at the same time from the same source, making them all cosmic cousins.[5]

To explain the differences between comets and asteroids as those terms are defined by the astronomers, we will refer to the proceedings of the NASA Detection Workshop:

> Asteroids and comets are distinguished by astronomers on the basis of their telescopic appearance. If the object is star-like in appearance, it is called an asteroid. If it has a visible atmosphere or tail, it is a comet. This distinction reflects in part a difference in composition: asteroids are generally rocky or metallic objects without atmospheres, whereas comets are composed in part of volatile substance (like water or ice) that evaporate when heated to produce a tenuous and transient atmosphere.[6]

Astronomers have long classified comets into three groups according to the length of their orbital period; that is, based on how long they take between each visit to the inner solar system. Comets fall into one of three classes: long period, short period, and intermediate period. The division among these groups is arbitrary, but it is generally accepted that any comet that revisits the inner solar system in less than twenty years is a short-period comet. This means that an astronomer has an opportunity to see the same comet twice in his professional lifetime. Long-period comets have periods of thousands to millions of years. The intermediate-period comets are defined as those with periods of more than twenty but less than two hundred years. The reason for adopting this definition is not clear, but it seems to be related to the fact that the scientific study of comets did not really begin until the most famous comet reappeared in 1758 according to a prediction made by Edmond Halley in 1705.[7] Halley's comet is still the best known, and has been studied more extensively than any other.

The vast bulk of the comets are thought by astronomers to lie in a region that has been named the Oort Cloud, after a renowned Dutch astronomer who first proposed the concept that there is a cloud of small bodies which he believed orbited the sun beyond the orbit of Pluto. This cloud, he thought, acted as a reservoir of comets. These will be discussed first. Comets located in the Oort Cloud were formed so far away from the sun that the only materials available for "construction" were mainly light gases, such as water vapor, carbon oxides, cyanogen, and the like. Thus, composition of comets is believed to be substantially different than that of asteroids, which formed just beyond the last "rocky" planet, Mars.

Oort comets are so far away from the sun that if anything disturbs their orbit and causes them to fall toward the sun, the fall requires a half billion years. These are, of course, all long-period comets. Hence, we need only to worry about any that are already falling toward us, even though we will not be able to see them for centuries. Of course, we have no idea how many there may be. We *do* know that several new comets are discovered each year; "new" in this case meaning comets never before noticed. Once the orbits of the new comets are measured, it is found that virtually all are of the long-period variety.

What causes comets to leave the Oort cloud and fall in toward the sun, and humans on Earth, is not really known. Several theories are popular among astronomers. It is known that the sun, together with its retinue of planets, is in an orbit around the center of the galaxy. Each trip around the galaxy takes about ten million years. One theory about Oort comets is that every so often, as the sun moves on its orbit, it has reasonably close encounters with other stars. The gravitational pull of these stars "stir" the Oort cloud enough that some comets are shaken loose from their rather unstable courses and sent tumbling toward the sun far below. Another theory is that the sun encounters occasional dust clouds in interstellar space, and that these disturb the comets and send them toward us. Still another theory holds that the gravitational field of the galaxy is "lumpy," just as the Earth's gravity field is slightly irregular, and that these gravity lumps rattle the comets free from their station in the Oort cloud.

Which of these theories, if any, is correct, is not of concern to us here. What matters is that we do know from looking through our telescopes that comets do come sailing in from the far edges of the solar system toward a collision with Earth. We do not even know how many comets are out there, although astronomers have guessed that there are a thousand million million (ten to the twelfth power) of them. That is enough to keep us supplied with surprises for many eons into the future. Running out of comets is not going to be one of our problems for a *very long* time.

There is a rather new theory, proposed by Gerald P. Kuiper, that there is yet another band of comets much closer to us than the Oort comets. These comets, located between the orbits of Saturn and Pluto, have rather stable orbits around the sun. Comets falling from this band would take a little less than a hundred years to impact the Earth, since they are so much closer. This would classify them as intermediate-period comets, as they have periods of less than two hundred years. In 1995 the Hubble Space Telescope took pictures of seventeen blurs that seem to confirm the existence of a "Kuiper belt" of comets, but astronomers are still discussing the existence of this group of comets. The frightening thing about these comets is that they appear to be vastly larger, and potentially far more deadly, than the Oort cloud comets. However, they are in much more stable orbits, and so are less apt to fall from the sky.

For many years astronomers have debated among themselves about the nature of comets. For some time the popular theory held that they were piles of rubble with various ices on the outside. As they approached the sun from their haunts in deep space, the sunlight warmed them and caused some of these ices to vaporize and form what is known as a coma. In 1951, Dr. Fred Whipple of the Smithsonian Astrophysical Observatory formulated a theory that comets are "dirty snowballs." This theory seems to fit all of the observations better than the "pile of rubble," and is now generally accepted as the preferred description of the comet nucleus. In this theory, comets are imagined to be a single lump of ice in a rocky matrix. Carl Sagan offers this description of some meteors that have been recovered:

> Those bright meteors that arrive from beyond Jupiter have been given the stirring name of transjovian fireballs. As determined from their entry characteristics, they are as fragile as the most delicate meteor known. If a sizable piece of such material were gently placed on a table before you it would collapse under its own weight. It is possible that the spaces in these silicate dust balls were originally on the parent comet, filled with ices and organics.[8]

A more recent theory suggests that comets may be a conglomeration of dirty snowballs, stuck together rather like a snowman. This theory gained credence with the comet known as Shoemaker-Levy 9, since it split apart before hitting Jupiter, apparently only from the influence of tidal gradients, which are a rather weak force.[9] The theory is supported by the evidence of strings of craters on Jupiter's moons. These craters, all about the same size, are nicely lined up like the results of a machine gun burst. They suggest that the impact of Shoemaker-Levy 9 is not unique, that similar events have happened before to Jupiter and its moons.

The proportions of ice and rock in these "dirty snowballs" is still a topic of discussion among astronomers. For many years it has been believed that comets were about 75 percent ice by weight and 25 percent the "dirt" in the snowball. Recent observations by our newest satellites, however, are raising a question about this ratio.[10] The Infrared Astronomical Satellite (IRAS) looks at cosmic objects by examining infrared light (invisible to the human eye) instead of visible light. Mark V. Stykes of the University of Arizona in Tucson found dust trails from the comets, especially as they neared the sun, that were brighter (that is, they had more dust) than anyone had suspected. From this Stykes deduces that the mixture of rock to ice is quite different than previously thought, with a much higher percentage rock in the mixture. This finding supports the hypothesis of some astronomers that the planet Pluto and Triton, Neptune's largest moon, were formed from an aggregation of comets, since these have the same density as that now postulated by Sykes for comets.

If this hypothesis is proven, the increased mineral content of comets will further increase the possibility that the dinosaur-killer of 65 million years ago might have been a comet. This would imply that comets are a bigger threat than presently believed, increasing the need to intensify our efforts at finding long-period comets.

This discovery may also show that comets do not last as long as previously thought. It had been believed that comets existed about ten thousand years before all of the gases in them had escaped and they, in effect, burned themselves out. The increased amount of dust may indicate that comets only last about five thousand years before extinguishing. (The oldest one we know about is Halley, and it has been observed for about 2,500 years.) This means that there may be a lot more extinguished comets masquerading as asteroids than anyone had suspected.

It should be noted that not all of the ice in comets is frozen water. Many other materials, most notably carbon monoxide and carbon dioxide, can take the form of ice. (The latter material is well known on Earth as "dry ice.") The material will boil off at a much lower temperature than frozen water, so a new comet, approaching from the frigid depths of the outer solar system, will start to boil away long before it reaches a temperature where frozen water would do so. Hence, we can often see comet-like features developing as far away from the sun as Jupiter.

Halley's comet was visited by three spacecraft as it came our way in 1986. Two were Russian probes, *VeGa 1* and *2*, and a third, *Giotto*, was launched by the European Space Agency. *Giotto* came closest to the nucleus of the comet, passing less than 400 miles from it. Many astronomers considered this to be a "suicide" mission, expecting little *Giotto* to be destroyed by dust and pebble impacts. *Giotto* did take some hits, but survived to take pictures and made measurements unprecedented in the history of comets.

We will briefly describe what was learned from *Giotto* and how it affects our prospects for dealing with future comets. First, we learned that the dirty snowball theory needs some refinement. There is a crust on the outside of the snowball. Ice does not boil off from all over the surface of the nucleus of the comet, as the simple snowball theory would suggest; rather, it comes out in discrete jets from a few places on the comet. The force of these jets can be calculated from the measurements made by *Giotto* as it flew past Halley.[11]

The various jets on Halley exert an estimated force of about five million pounds, or nearly as much as all of the engines of the space shuttle as it lifts off from the launch pad.* And these jets continue for hour after hour, day after day, waxing and waning somewhat over time. They are not all pushing in the same direction, however. Since the comet has an estimated mass of perhaps 80 billion tons, the jets do not affect the comet's orbit very much.[12] Further, since the comet rotates, the jet effect is largely

---

*This is a very simple calculation, although the numbers are very approximate. According to the reports, Halley is venting about twenty-five tons of gas per second at a velocity of 900 meters per second. Hence, from the rocket power equation, the thrust is 25,000 kilograms per second times 900 meters per second, o4 22.5 million Newtons (5.06 million force pounds).

averaged out, except for the fact that the force varies with time. The lesson for us is that comets will need to be watched very closely, since they have the ability to zig-zag toward us unexpectedly.

Second, the pictures of Halley's nucleus indicate that it is bigger than expected. The nucleus is much blacker than believed; it is as black as soot. It is about ten miles long, five miles wide, and five miles in depth. The pictures taken do not show much detail, since there was so much dust around the nucleus, but there are good indications that the comet's surface is bumpy, as if it has mountains and craters. However, the crust is believed to be quite fragile, which would make it hard to attach anchors if ever we wanted to tow it away from an Earth-intercept path.

When the surface temperature of Halley was measured, spots approximately the temperature of boiling water were discovered, but tens of feet down the comet may still be quite cold, since the surface crust is expected to be a very good insulator. We do know that Halley contains a *lot* of water, and presumably other comets do as well. The gas surrounding Halley was 80 percent water vapor, with carbon monoxide and carbon dioxide constituting most of the balance.

The latest sensation in the comet world is one named Hale-Bopp, after the men who detected it in July 1995. It was discovered unusually far from the sun, which suggests that it is big. Present estimates, based upon observations from the Hubble Space Telescope, are that it is several times larger than Halley. Although its size does not necessarily mean that it will be brighter than Halley, astronomers expect that to happen. Comet Hale-Bopp will be closest to the sun in April 1997, and there is a good chance that it will become visible to the naked eye in 1996.

This chapter has detailed the various theories regarding the origins of both asteroids and comets. Asteroids appear to be the result of a planet which either failed to form properly or formed and then exploded from the force exerted by Jupiter's gravitational field. The existence of asteroid families seems to support this theory.

Meteorites, many of which have been found in the Antarctic, are pieces of asteroids and are smaller than the asteroids themselves. Studies have proven that some meteorites which have been found originated on the moon or Mars. It has recently been discovered (in August 1996) that at least one Martian meteorite contains evidence for what may have been life forms that existed billions of years before there was any life on Earth.

While asteroids are primarily rocky in their composition, comets can be described as "dirty snowballs," i.e., as a single lump of ice in a rocky matrix. Comets contain a large proportion of a frozen substance (such as water, carbon dioxide, or carbon monoxide) beneath a rocky crust. As the comets near the sun, this frozen matter evaporates, producing a transient atmosphere. The space probe *Giotto,* which flew close to the nucleus of Halley's Comet, transmitted information indicating that there are openings on the comets' surfaces through which the evaporating gases are released in jets.

The next chapter explores a theory regarding the main cause of the extinction of the dinosaurs some 65 million years ago. A father and son team, Luis and Walter Alvarez, first offered their hypothesis involving the crash of a giant asteroid on Earth a decade and a half ago. Subsequent findings have corroborated the Alvarez team's discovery, allowing the hypothesis to withstand scientific scrutiny. It is now accepted by most scientists as fact.

## Notes

1. Dan Durda, "All in the Family," *Astronomy* 21, no. 2 (February 1993).
2. Leon Jaroff, "Life on Mars," *Time* (August 18, 1996): 60.
3. John Noble Wilford, "Signs of Primitive Life on Mars are Found in Ancient Meteorite," *New York Times*, 7 August 1996, p. 1.
4. "On to Mars," *Time* (August 18, 1996): 12.
5. Thomas Van Flandern, "The Origin of Sixty Comets," *U.S. Naval Institute Proceedings* (April 1976): 18.
6. *Report of the NASA International Near-Earth-Object Detection Workshop* (Washington, D.C.: GPO, 1992).
7. Carl Sagan and Ann Druyan, *Comet* (New York: Random House, 1991).
8. Ibid.
9. Ron Cowen, "Are Dirty Snowballs Made of Smaller Ones?" *Science News* 145 (May 7, 1994): 298-99.
10. Ron Cowen, "Comets: Mudballs of the Solar System?" *Science News* 141 (March 14, 1992): 170-71.
11. Richard Berry and Richard Talcott, "What Have We Learned from Comet Halley?" *Astronomy* 14, no. 9 (September 1996): 6–22.
12. Richard Berry, "Halley in Heidelberg, Planets in Paris," *Astronomy* 15, no. 1 (January 1987): 24–32.

# 5

# Did an Asteroid Kill the Dinosaurs?

These characteristic features of shocked quartz at several sites world-wide confirm that an impact event at the Cretaceous-Tertiary boundary distributed ejecta products in an Earth-girdling dust cloud, as postulated by the Alvarez (asteroid) impact hypothesis.
—Report of Researchers from the U.S. Geological Survey,
*Science,* May 9, 1987

## The Strange Limestone Findings in Gubbio

In the 1970s, Dr. Walter Alvarez, a geologist of the University of California at Berkeley, conducted some geological digs in Gubbio, Italy, which is about seventy miles southeast of Florence. While there, he discovered a reddish brown stripe in a limestone cliff in the Apennine Mountains in Italy. When this clay was analyzed, it was found to have thirty times the normal amount of iridium, an element rare in the Earth's crust.[1] (Iridium is a precious metal common in meteorites.) He discussed his finding with his father, Luis Alvarez, a famed nuclear physicist. Father and son talked about the oddity of the thin brown line. Searching for a

76

project that they might work on together, Walter and Luis decided to pursue the consequences of their findings as a team.

This thin layer of reddish brown clay was at the very boundary of geologic time between two periods, the Cretaceous Age (the age of dinosaurs) and the Tertiary age ( the age of mammals).* From the initial discovery of the clay, they eventually concluded that there had been a gargantuan meteor impact at the time this layer of earth was deposited, resulting in the extinction of two-thirds of all species living at the time, including the most famous, the dinosaurs.

Iridium was selected for study (rather than platinum, for example) because it can easily be detected in minute quantities. The men used a nuclear-age technique for geochemical analysis, namely, neutron activation analysis. This method uses a small research reactor to irradiate the sample. The iridium readily absorbs the neutrons, so that after a brief exposure the characteristic radiation can be measured and the amount of iridium quite can be determined accurately.

Tests of clay from the K-T boundary at several locations around the globe showed that iridium was present in amounts completely out of proportion to the clay above and below this geological boundary, which has been determined to be 65 million years old. The Alvarez team postulated several theories to account for the vast quantity of iridium that was represented, and concluded that the only plausible theory was an asteroid impact. They even calculated the size of the asteroid as being from four to eight miles wide.

The paleontological community reacted with fury to this theory, believing that a geologist and a nuclear physicist had inadequate knowledge of the dinosaurs to determine what had led to their extinction. This chapter recounts of that controversy and where it stands today.

## A Challenge to the Asteroid Extinction Theory

The first serious scientific objection to the Alvarez theory was the absence of a giant crater on the Earth which would show that a large extraterrestrial body had hit the planet. If the asteroid had been as large as

---

*This is called the K-T boundary in scientific literature so as not to cause confusion with the Cambrian period, which is abbreviated as "C."

the theory required, it would have made a crater at *least* a hundred miles wide and probably bigger. The only large craters known at the time the Alvarez theory was proposed were of the wrong age to be the scar left by the asteroid postulated to have caused the extinctions. The absence of a crater, however, can be explained by the fact that a crater 65 million years old would have been badly eroded by wind and water over the eons. Another explanation is that it might be under water, hidden deep beneath the oceans.

The paleontologists took every advantage of the missing crater to advance their own ideas about the extinction of the dinosaurs. Three years after the asteroid hypothesis was formulated to explain the mysterious death of the dinosaurs, two American scientists, Michael Rampino of NASA's Institute for Space Studies at New York University and Robert Reynolds of Dartmouth College, reported some new geological evidence that challenged that theory. Rampino and Reynolds analyzed deposits from four different geological sites (in Tunisia, Spain, Italy, and Denmark) which marked the transition of the Earth from the Cretaceous to the Tertiary Period. As a result of their study, the two men devised an alternative explanation for the mass extinctions based upon global volcanic activity.

They described their findings in the February 4, 1987, issue of the academic journal *Science*. Rampino and Reynolds reasoned that there should be a striking uniformity in the clays from all the localities, whereas Alvarez discovered iridium concentrations varying from 20 to 160 times higher than normal in the thin beds of clay marking the end of the period. Their conclusion was that the clay, reflecting the composition of the material thrown up from the point of impact, was "neither mineralogically exotic nor distinct from locally derived clays above and below the boundary."[2] In conclusion, the NASA/Dartmouth team opted for an immense volcanic activity at the end of the Cretaceous Period with the resulting lethal dust cloud and disruption of the food chain as the main reason for the extinction of the dinosaurs, a finding in direct contradiction to the analysis made by the Alvarez team, which explored a very similar possibility.

Other skeptics also claimed that the dinosaur extinctions were probably earthly (i.e., volcanic) rather than extraterrestrial in their origin. Several scientists theorized that the iridium layer was deposited over a period of up to 100,000 years and could have come from large volcanic eruptions originating deep within the Earth. The dinosaur extinctions, according to Drs. Charles Officer and Charles Drake, both of Dartmouth College, occurred around the times that a huge eruption in what is now India created a great lava field known as the Deccan Traps.

Other experts generalized about the volcanic idea, finding that nine of the ten greatest mass extinctions—including that of the dinosaurs—more or less coincided with enormous floods of lava in various parts of the world. They knew that strong earthquakes sent out shock waves that propagated through the Earth and focused at the quake's antipode (i.e., at the point of the Earth directly opposite the quake's epicenter).

## But, What Killed the Dinosaurs?

The dinosaurs were one of the most successful species ever to live on Earth. They populated the planet for 130 million years, from 195 million B.C.E. to 65 million B.C.E. This period of domination is longer than for any other known species. Whatever happened to these magnificent creatures that children today find so fascinating?

The Alvarez team felt that the explosion resulting from the asteroid impact could have thrown up enough debris and dust to blot out the sun and bring about the extinction of most life forms on the planet. Their argument was that the sunlight had been so diminished by the dust that all photosynthesis had stopped and the food chain was cut off, starting at the lowest level, microscopic floating plants (algae), and proceeding all the way up to land-based plants that animals eat. Since this condition existed for several years, animals starved to death. The Alvarez team had found the iridium deposited just on top of the highest, and therefore the most recent, stratum of rock contemporary with those bearing dinosaur fossils. Deposits above and below the clay boundary layer separating the Cretaceous layer from the succeeding Tertiary have very little iridium. (Since 1980, iridium anomalies have been found in more than 80 different places around the world, including deep sea cores, and in layers of sediment formed at the same time, which lends support to the Alvarez theory.)

In 1988, Rampino and two other New York University-NASA colleagues published an article in the British journal, *Nature,* theorizing that the dinosaurs probably roasted to death because 95 percent of the marine algae disappeared at the same time the dinosaurs did, and because of the dust layer that enveloped the Earth after the meteor impact. Rampino deduced that with few algae left in the oceans, and few clouds in the sky to filter out the sun, the hot solar rays would have caused the temperatures to rise by eleven degrees Fahrenheit. Thus it was the excess heat and lack of food, not the dust clouds alone, that brought about the end of the dinosaurs, since they were unable to cope with the impact of the higher temperatures on their bodies.

## A Second Alternative

After the volcanic theory was used to explain for the extinction of the dinosaurs, another theory was put forth by Edward Anders of the University of Chicago. He had been studying dinosaur fossils unearthed by William Clemens of the University of California on Alaska's North Slope (where oil was discovered). Anders felt that the dinosaurs may have escaped the dust-choking effects of an asteroid impact.

Anders found vast amounts of soot in the layer of rock surrounding the dinosaur bones, and like samples which had been brought in from Denmark and Australia, he hypothesized that the soot probably came from vast wildfires. Anders believed that these wildfires were probably started from heat caused by an expanding fireball of incandescent hot rock vapor that was given off when the asteroid burned its way through the atmosphere. As the vapor cooled, the scattered hot rock particles would still have been hot enough to start many wildfires. Anders reasoned that if the dinosaurs were living that far north, that they must have adapted to the long dark winters, and that the darkness caused by dust would not have killed them. Neither could they freeze, since the nearby Arctic Ocean was warm in those semitropical days, modifying the climate. The theory is weak in that it does not explain what the dinosaurs ate for several years while there were few plants living at the bottom of the food chain.

## The Crater Is Found...At Last!

The biggest objection to the Alvarez theory was removed in the 1980s when a giant crater was discovered in the Gulf of Mexico. In 1978, consulting geologist Glen Penfield was searching for promising oil exploration sites near Chicxulub, on the northern tip of the Yucatan peninsula in Mexico. As he pieced together bits of geological information, Penfield began to uncover a pattern.[3] From magnetic and gravity anomaly data, he found what appeared to be a giant crater. Because Penfield's data belonged to a Mexican oil company, these geological findings were not published until 1981. A fraction of the rim of the crater is on land, and it is mostly hidden by erosion, forests, and the carbonate (coal and oil) sediments. The balance of the crater lies under water (and under sediments) and can only be detected by scientific instruments.

In order to establish what the discovery meant, more research was necessary after the crater was first detected. University of Arizona graduate student Alan R. Hildebrand had been looking for the K-T (Alvarez) crater in Haiti. He concluded from what he found there that the crater he sought must be within a thousand miles of Haiti. When he heard about Penfield's findings, they teamed up, and by searching through old drilling core had established by 1991 that there was a very high probability that this was, indeed, the long-sought site of the K-T impact. Further research determined the crater to be of about the correct age to be the "smoking gun" for the Alvarez theory, thereby enhancing the claim.

Penfield's find was identified as a multi-ring crater.* The third ring was initially believed to be the outer (largest) and it was about 120 miles in diameter. This was not quite big enough to satisfy the paleontologists, who wanted an even bigger crater. The size of the asteroid that Alvarez had predicted to satisfy his theory was about six miles wide. Hypervelocity impact studies predict a crater to be about twenty times the diameter of the intruding asteroid, therefore, Penfield's discovery was credible as the impact site, but just barely.[4]

The problem of size regarding Penfield's crater was eliminated in 1993, when a team headed by Virgil Sharpton did a new study of gravitational anomalies in the Chicxulub region.[5] Sharpton's team published an article in *Science* that showed the size of the various rings of the crater that was formed 65 million years ago. Penfield was correct in estimating

---

*A multi-ring crater is formed when a hypervelocity object (one traveling at a very high rate of speed) hits a larger solid. When a stone falls into a pond, rings form around the splash point. When a solid mass, such as the Earth, is hit by a large, fast-moving meteor, it behaves the same way, except that it solidifies quickly, freezing several of the rings into stone.

the third ring at about 120 miles in diameter, but there is a *fourth* ring, which measures about 185 miles in diameter. This is big enough to embrace the asteroid required to supply all of the iridium found worldwide at the K-T boundary.

The lead author of the study, Dr. Virgil L. Sharpton, of the Lunar and Planetary Institute in Houston, noted

> It's hard to conceive all that energy being released, vaporizing thousands of cubic miles of the Earth. This latest study added support to the [1980] previously published idea that a "doomsday rock" from outer space did, indeed, end most life on the planet by creating a pall of dust that blotted out the sun, as well as bringing on firestorms and acid rain. This study proved that the Yucatan crater is the largest known celestial blemish to mar the Earth's surface.[6]

The two closest rivals to Penfield's crater on Earth are located in Sudbury, Ontario, and Vredefort, South Africa. The diameter of these two craters has been computed at no more than 125 miles across. The closest extraterrestrial rival to the Yucatan crater is the Mead Crater on Venus, which is about 175 miles across. It is named after American anthropologist Margaret Mead, who died in 1978.

The research team under Dr. Sharpton, including several Mexican scientists, analyzed nearly 7,000 readings of gravitational strength that had been made over the suspect Yucatan region to come to their conclusions regarding size. The Yucatan crater scientists detected three major rings and parts of a fourth, expanding in ever increasing circles, not unlike the pattern of a bull's eye. The spacing of the rings was similar to that found on big craters on other planets in our solar system.

"Earth," Sharpton and his associates concluded, "probably has not experienced another impact of this magnitude since the development of multicellular life approximately a billion years ago." They added that an event of such power would have produced environmental havoc and that its occurrence exactly at the end of the reign of the dinosaurs (65 million years ago) "argues strongly that the effects of this meteorite impact led to the concurrent mass extinction."[7]

Meanwhile the clock is ticking on the countdown to the next asteroid hit. The only questions remaining are *when* and *where*.

Seven years after the Alvarez hypothesis, in the May 1987 issue of *Science*, scientists Bruce Bohor, Peter Modreski, and Eugene Foord, researchers at the U.S. Geological Survey, reported that they found shocked quartz crystals in eight widely separated places, ranging from Europe to New Zealand, indicating a global phenomenon. The crystals were embedded in the same sediment layers that contained abnormally high amounts of iridium. The survey team concluded that the fracture patterns in the quartz ruled out large-scale volcanic eruptions as a cause of the extinctions.

## New Multiple Jeopardy Extinction Theory Evolves

As the 1990s began, two main schools of thought clashed furiously over the controversial question of what brought about the dinosaurs' demise. One school held that a massive (six-mile-wide) object—most probably an asteroid—slammed into the Earth from outer space, off the Yucatan Peninsula, kicking up a worldwide pall of dust. The sun was blotted out and most of the plants and animals were exterminated.

The other school held that the main cause of the global mayhem was natural Earth processes such as major volcanic eruptions. By the end of 1994, however, an elegant new theory emerged combining both conflicting ideas into a single explanation.[8]

This new theory proposed that since a speeding asteroid colliding with the Earth would unleash a force equal to millions of hydrogen bombs, it would send gargantuan shock waves through the Earth. These waves would have coalesced on the opposite side of the impact crater in an area known as the antipode, where ground would be broken and dozens of volcanoes would erupt. The impact and its repercussions in this opposite hemisphere could have contributed to the death of the dinosaurs some 65 million years ago.

This theory, known as *antipodal vulcanism*, first surfaced in the early 1990s and soon took on weight as computer modeling began to suggest its plausibility. A team of scientists from the Sandia National Laboratory in Albuquerque, New Mexico, used a powerful computer in the fall of 1994 to simulate the effects a speeding asteroid some six miles in diameter—the estimated size of the dinosaur-killer—would have had at the impact's antipode. They discovered that the crust at that spot would have heaved as high as sixty feet in a series of catastrophic tremors. (By comparison, the ground around the great San Francisco earthquake of 1906 moved only a few feet at most.)

"The Earth acts as a lens," said Dr. Mark Boslough, a Sandia physicist, who lead the simulation effort. "It focuses the energy. There has been a lot of speculation about this in relation to asteroid impacts and volcanic eruptions, but we've done the first vigorous modeling to show where the energy actually goes."[9]

With such evidence in hand, Dr. Hagstrum of the Geological Survey and a colleague, Dr. Brent Turrin, published an outline of the antipodal extinction idea in 1991, suggesting that giant bombardments over the eons had touched off heavy volcanic flows.[10] These, linked with the upheavals, caused the mass extinctions. But evidence was hard to assemble due to the weathering of the Earth's surface over the ages.

Some paleontologists, still clinging fervently to cherished ideas, insist that the cause of the dinosaur extinction was volcanic eruption, citing the Deccan Traps in India as the proximate cause of that event. The asteroid impact theory, however, also suggests that the surface of the Earth would be fractured at the antipodes of the strike, thereby triggering volcanic eruptions. Doubters of the Alvarez hypothesis point out though, that the Deccan Traps in India are *not* at the antipode of the Chicxulub crater.

We think there is a simple explanation for this. Pockets of molten lava (called "plumes") are believed to lie close to the surface of the Earth, the result of molten material deep in the Earth rising and trying to find an outlet.[11] Such plumes, close to the surface, would cause a weak spot in the crust of the Earth. Engineers know that an object does not always break where the stress is highest, but will often do so at a nearby point, where the material is weak and the stress is also high. The Deccan Traps are close enough to the antipodes of Chicxulub that the fracture that released all the lava could easily have been caused by the stress resulting from an asteroid impact on the other side of the Earth.

Dr. Hagstrum has a still different theory. He suggests that 65 million years ago, during the great dying off of the dinosaurs, the Deccan Traps of India were antipodal not to the Yucatan Peninsula, where the great crater lies, but to a spot in the eastern Pacific Ocean . . . where the seabed bears some evidence of a long ago major impact.

The theorizing was joined in 1992 by Dr. Michael Rampino and Dr. Ken Caldeira, a geologist at Penn State University. They published a paper that concluded that hot spots within the Earth were responsible for enormous lava floods that tended to occur in antipodal pairs, which might have been triggered by cosmic asteroid bombardments.

The most recent work in this field was presented by Mark Boslough and the Sandia group at an October 1994 scientific symposium held in Santa Fe, New Mexico. Scientists at the Sandia National Laboratory postulated that a six-mile-wide asteroid, hitting the Earth at 45,000 miles per hour, would have dug out a colossal crater that a few seconds after impact would have measured more than fifteen miles deep. Top pressures generated by the impacting asteroid would have been about six million times greater than the force exerted by the Earth's atmosphere. Eighty minutes later, the shock waves would have rippled through the Earth via various catastrophic routes to the antipode, where they would have converged in a series of giant shock waves that should have shaken the upper one hundred miles of the Earth's crust, thus creating a pipeline of destruction from the depths to the surface.

Dr. Boslough confirmed that it would take the supercomputers at Sandia (an arm of the Department of Energy) until the end of 1995 to make the heat calculations.[12] The results of that analysis were not yet available as this book went to press.

## A Unified Theory of Dinosaur Extinction

We now have in hand all of the pieces we need to draw a fairly coherent picture of what might have happened 65 million years ago. A giant Cosmic Object Threatening Earth (COTE) slammed into the planet with an explosive force beyond imagination. At the moment of impact it made a dent in the Gulf of Mexico that was twenty-five to thirty-five *miles* deep, which subsided to a crater only ten miles deep. In other words, the Earth's surface reacted with an elasticity not usually attributed to it and sprang back nearly into its original place. Ten miles straight down, excavated land hurled into the sky. There is a large sulphur deposit at that location in the Gulf, so many billions of tons of sulphur were hurled into the sky, along with many more billions of tons of rock and dust. These rocks were thrown thousands of miles into the sky, far beyond the atmosphere. (As a matter of fact, some of the COTEs probably hit the moon. We do know that bits of the moon have been knocked to Earth by celestial impacts.) The rocks from the Gulf were thrown so high that some of them took hours to fall back to Earth. Meanwhile, the Earth continued to rotate, ringing like a sharply struck gong, but at a very low frequency. As the bits of rock came crashing down, they must have been falling so very fast (approaching escape velocity) that they acted like a very intense meteor shower. The sky was alive with fire. As occurred with the impact at Tunguska in 1908 (discussed in chapter 1), the vegetation all over the planet burst into flames.

At the antipodes, the Earth burst, spewing out cubic miles of molten lava. This started about an hour after the impact, and long before most of the rock had yet fallen back to Earth. Earthquakes from the hammer blow that had been struck shook the ground everywhere. In the sky, clouds of dust began to spread to cover the sun. Based upon the results of atomic bomb tests, it was believed that it took months before the dust concealed the sun everywhere, but it soon grew too dark to support plant life and most of the inhabitants of the planet died. The plants died first, and then the herbivores that hadn't already been killed by fire, falling stones, earthquakes, or molten lava, starved to death. In a matter of days or weeks, millions of tons of sulphuric acid began to rain down on the hapless creatures still living. This deadly rain, as potent as battery acid, polluted waters everywhere, killing most of the few remaining plants and animals.

## The Paradigm Theory of Mass Extinctions

There is a growing conviction among many scientists, however, that cataclysmic simultaneous extinction of many species has been a recurrent phenomena that plays a far more important role than almost anyone had hitherto suspected. There is some evidence from fossil records that extinctions have occurred at regular intervals of approximately 26 million years. Many people are coming to believe that these also result from extraterrestrial impacts, a belief which has already gained enough support to force a reappraisal of the basic evolutionary theory in a fundamental transformation of thought known as a "paradigm shift."

There is one school of biologists today that believes that the asteroids over the ages have periodically ended the rule of the dominant living species of the time, thus encouraging new life forms to take over the ecological gaps, and to shape the evolution of life on Earth. A major paper was published in 1984 by David Raup and J. John Sepkoski, paleontologists from the University of Chicago.[13] Based on the fossil record of extinctions over the last 600 million years, they found major planet-wide extinctions occurring on a regular basis, about every 26 million years.

There are several theories to explain why these mass extinctions occur regularly. One thesis, proposed in 1984 by Marc Davis and Richard Muller of the University of California at Berkeley, and Piet Hut of the Institute of Advanced Study at Princeton, postulates a companion star to the sun on a 26-million-year orbit. As it returns close to the sun, this star disturbs the Oort cloud located out beyond Pluto and sends many comets hurling our way, perchance to crash into Earth and annihilate many of its species. This companion star has been named Nemesis after the Greek goddess of vengeance.

This theory is not without problems, however. A star with a period that long would have its farthest retreat from the sun (aphelion) a third of the way to the nearest star. Such an orbit would not be very stable, and it is doubtful if it could have stayed in such an orbit for more than a fraction of the life of the solar system.

It is known that the sun's passage through the universe wobbles above and below the galactic plane, with a half period of about 32 million years, give or take a million years. So every 32 million years, we are in the plane of the galaxy. Could the galaxy's gravitational pull affect the Oort Cloud and bring down a rain of comets every so often? A couple of scientists from the NASA-Goddard Institute for Space Studies re-examined the extinction data to see if this fit their theory. Not too surprisingly, they came to believe that a period of from 31 to 33 million years gave a better fit to the extinction data than 26 million years.

"How can the data fit both 26 million and 32 million years?" came the question. To confuse the issue still more, another analysis by Walter Alvarez, who had originally formulated the asteroid theory, and a colleague looked at the history of identified impact craters on the Earth, and found that periodically, the Earth had been hit by asteroids—but the interval was 28.4 million years, not 26 or even 32 million! So there may, or there may *not*, be a periodicity in extinctions.

If there is such a periodicity, there is at least one other theory that tries to explain it. This postulates a tenth planet. If a tenth planet exists, this theory holds, it is in a highly eccentric inclined orbit compared to the orbits of the other nine planets, sweeping from near the sun to out near the Oort cloud. Periodically, its precession moves it to a position where it can disturb the Oort cloud and thereby send comets hurtling toward Earth.

But, all of these theories about periodicity involve comets, not asteroids hitting the Earth. Is is possible that it was a comet that impacted the Earth? The tentative answer appears to be yes. It is conceivable. Scientists are reasonably sure it would have had to have been large to possess enough mass to cause the crater which has been observed. The other thing that is known is that meteorites, which are presumed to come mostly from asteroids, contain large amounts of iridium. But if some of these meteorites came from comets, that would imply that comets are also high in iridium. So, it is possible it was a comet and not an asteroid that killed the dinosaurs. In either case, there are a lot more cosmic objects out there that threaten Earth, and we need to mount a watch to look for them.

## There Was a Great Crash

Virtually every scientist now agrees that 65 million years ago there was a major impact of a cosmic body with the Earth. The evidence is too overwhelming for anyone to still doubt it. The remaining questions are:

1. Was that cosmic object a comet or an asteroid?
2. Exactly what mechanism killed the dinosaurs? and
3. Why did some life survive?

Not all of the paleontologists, however, have given up the battle. Some still contend it had to be something other than an asteroid impact that killed the dinosaurs. For example, William A. Clemens and L. Gayle Nelms of the University of California, Berkeley, still argue against the Alvarez hypothesis, even in the face of all of the evidence. They discovered dinosaurs they believe to have wintered in the Arctic, therefore, they claim the cold and darkness subsequent to an asteroid impact would not have killed at least those dinosaurs.[14] One of their supporters, Tom Rich of the University of Monash in Australia, said the impact theory as an explanation for the extinction of the dinosaurs, was "simplistic." The creatures would not have died simply because a comet "turn[ed] off the lights for three months."[15] Obviously Rich has never read the original Alvarez paper, which pointed out, based upon the experience of Krakatoa which was previously described, that the sun would have been dimmed for several *years* by the postulated impact.

The original paper by the Alvarez team assumed that the dust cloud was spread by atmospheric winds, just as the dust from the volcano eruption of Krakatoa was, and that it took as long to settle out. Now it appears that the explosion was so incredibly powerful that much of the dirt and rock kicked up by the impact was spread much more rapidly than that; much of it went into space on ballistic trajectories and came down all over the Earth in a matter of hours.

The volume of rock and dust thrown into the atmosphere by the eruption at Krakatoa is estimated to have been eighteen cubic kilometers (50 billion tons).[16] That eruption did not begin to obscure the sun, but it caused spectacular sunsets for two or three years following the event. The Alvarez paper estimated that the asteroid impact put a thousand times more dust into the atmosphere. That would definitely cut off the sun to such an extent that photosynthesis in plants would be stopped. Since this process at the bottom of the food chain is essential, the whole chain would collapse in short order. Although the atmosphere contained a thousand times more dust, it would not take a thousand times as long for the dust to settle, since it would settle faster at first. But it would certainly require a number of years before food all over the planet again became adequate.

How any animal could survive that long with the food chain disabled is a mystery:

> The real question is, how did the others—how did any animal—manage to survive? [Impact theorists] have got to come up with a hypothesis that puts equal weight on survival. So many of these catastrophists want to kill the dinosaurs, they forget the rest of the biota. Birds, mammals, and amphibians managed to survive, and that tells you that there is something wrong with most of these hypothetical horrors."[17]

Clemens poses a good question, and no doubt we shall find the answer in due course.

How much longer will this argument go on? We like the answer given by William Glen, a geoscientist and historian with the U.S. Geological Survey in Menlo Park, California, as quoted in *Science News*. He said, "In the case of an upheaval in science in which a new thesis or paradigm is offered up, the last vestiges of debate cease only with the death of the old guard."[18]

## The Shadow of Dr. Luis Alvarez

If the asteroid impact theory holds up, then it appears that the dinosaurs died because they had the bad luck to be roaming over the Earth at the time of the hit and not because they succumbed to some disease. Dinosaurs dominated the Earth for 160 million years and might have thrived up until today, but we wouldn't have been here to see them, since the event that closed the dinosaur era opened the way for the ascension of primates and other higher mammals.* Dinosaurs were well adapted to a largely tropical world, and when the world suddenly turned fiercely hostile, it killed them.

A clue to the eventual acceptance of the Alvarez-asteroid impact theory can be found in a commentary about the late physicist's approach to an unsolved scientific problem.

In a book about his work with Dr. Luis Alvarez (who won the coveted Nobel Prize in 1968 for his use of bubble chambers to detect new subatomic particles in matter), Dr. Richard Muller, a protege and fellow physicist, described his mentor's style as a physicist this way: "Alvarez's whole approach to physics was that of an entrepreneur, taking big risks by building large new projects in the hope of large rewards, although his pay was academic, rather than financial. He had drawn around him a group of young physicists anxious to try out the exciting ideas he was

---

*Many people believe that it has been mass extinctions which make possible the processes of creation through a continuing cycle of death and renewal. While a mass destruction annihilated the dinosaurs, it was an earlier mass extinction (or extinctions) occurring at the end of the Permian Period (225 million years ago) that closed the age of the amphibians and made the world available for the dinosaurs to arise and diversify.

proposing. . . . [He] seemed to care less about the way the picture in the puzzle would look when everything fit together, than about the fun of looking for pieces that fit."[19]

The Alvarez thesis on the asteroids lives on today as the most plausible explanation of the death of the dinosaurs—and as a warning to us to "look out there" for another possible "doomsday asteroid" that just might bring about the end of our human civilization as we know it.

In this chapter we have analyzed the scientific tug of war that has raged for over a decade since Luis Alvarez and his son, Walter, postulated the asteroid dinosaur-killer theory. Other theories focusing on volcanic eruptions have been proposed to account for the dinosaurs' demise, but as of this writing, the Alvarez hypothesis is the most accepted in scientific circles.

In the next chapter an analysis is made of the origin and content of the Morrison Committee's Spaceguard Survey, commissioned by NASA and published in January 1992. This first, small step in speeding up the search for, detection of, and ultimately the deflection and destruction of incoming asteroids has been long in coming and should lead to more such committees and ultimately action in the future.

## Notes

1. Luis W. Alvarez and Helen V. Michel, "Extraterrestrial Cause of the Cretaceous-Tertiary Extinction," *Science* 208 (June 6, 1980): 1095–1108.
2. Michael Rampino and Robert Reynolds, "Death of the Dinosaurs," *Science* 234 (February 4, 1987).
3. J. Kelly Beatty, "Killer Crater in the Yucatan," *Sky & Telescope* (July 1991): 38–40.
4. Eugene M. Shoemaker, "The Collision of Solid Bodies," in *The New Solar System*, J. Kelly Beatty, Brian O'Leary, and Andrew Chaikin, eds. (New York: Cambridge University Press, 1990), pp. 33–44.
5. Cited in Virgil L. Sharpton et al., "Chicxulub Multiring Impact Basin: Size and Other Characteristics Derived from Gravity Analysis," *Science* 261 (September 17, 1993): 1564–67.
6. William Broad, "New Clue to Cosmic Collision and Demise of the Dinosaurs," *New York Times*, 17 September 1993, pp. 1 and 14.
7. Sharpton et al., "Chicxulub Multiring Impact Basin."

92    DOOMSDAY ASTEROID

8. William Broad, "New Theory Would Reconcile View of Dinosaur Demise," *New York Times*, 27 December 1994, pp. C1 and C10.

9. Ibid., p. C10.

10. Ibid.

11. Dana Desonie, *Cosmic Collisions* (New York: Henry Holt & Co., 1995), pp. 76–77.

12. Broad, "New Clue to Cosmic Collision."

13. Paul R. Weissman, "Are Periodic Bombardments Real?" *Sky & Telescope* (March 1990): 266–70.

14. Virginia Morell, "How Lethal Was the K-T Impact?" *Science* 261 (September 17, 1993): 1518–19.

15. Ibid.

16. Alvarez and Michel, "Extraterrestrial Cause," p. 1105.

17. Morell, "How Lethal Was the K-T Impact?"

18. Richard Monastersky, "Impact Wars," *Science News* 145 (March 5, 1994): 156–57.

19. Ibid., p. 157.

# PART TWO

# THE SEARCH AND LANDING PROCESS

# 6

# Spaceguard Survey: Find the Celestial Objects Threatening Earth

In a recent discussion of the effort involved in the search and detection of asteroids, David Morrison, the Chief of Space Science at NASA's Ames Research Center, pointed out that: "the total number of people engaged in this sort of work worldwide is less than the staff of an average McDonald's restaurant."[1]

There is a story of an inebriated gentlemen hunting for something under a lamp on a street corner. A friend stops and tries to help. Apparently, the fellow has lost his keys and can't get into his house. After several minutes of fruitless searching, the friend asks the drunk:

"Are you sure this is where you dropped your keys?"

"Oh, no." the drunk replied. "I dropped them over there," he said, pointing down the street a hundred feet or so.

"Then why are you looking for them here?" the friend asked.

" 'Cause the light is so much better over here," came the answer.

The above anecdote is rather like the solution for finding asteroids proposed by a NASA committee in a report to Congress dated January 25, 1992. This committee was established in response to a section of a 1990 Congressional appropriations bill directing NASA to conduct two studies regarding the danger of an asteroid impact. The first study was undertaken to propose ways to find these cosmic objects before they hit us. The second was to to consider means to deflect or destroy the cosmic objects after detection. The work of this second committee will be discussed in later chapters.

The first committee was chaired by David Morrison, an astronomer from the NASA Ames Research Center, and was composed almost entirely of eminent astronomers. However, a review of the roster does not indicate that anyone who could provide engineering expertise in such disciplines as interplanetary trajectories, radar, lasers, or other relevant disciplines was included on the committee. This distinguished group met three times in 1991 at various locations to consider the problem they had been assigned, and issued their report in January 1992.

This group, henceforth referred to as the Morrison committee, reached four basic conclusions:

1. The major risk from which planet Earth must be protected is posed by asteroids nearly as large as that which killed off the dinosaurs 65 million years ago, rather than the much more frequent impacts of smaller objects, such as the one that devastated the Tunguska region of Siberia in 1908.

2. The search for these objects could be conducted by Earth-based telescopes alone, and the best strategy is to look into the midnight sky. (We will show, later in this chapter, however, that this is "looking where the light is good," rather than looking to where the threat is.)

3. The astronomers concluded that they needed six large new telescopes (costing about $50 million each) to conduct this search for threats from the sky. If they were given this, and $10 to 15 million per year for expenses, they believed they could find about 75 percent of the objects that were big enough to wipe out a third of all humans, and do it within about twenty-five years. The total program would cost at least $300 million.

4. The search for errant asteroids and wayward comets needed to be an international enterprise.

This last conclusion is the only one with which we can agree. This chapter will focus on why looking only with Earth-based telescopes and only near midnight is looking where the light is good and not a very good solution. Additional areas of disagreement are discussed in other parts of this book.

Astronomers usually draw their maps of the solar system on a diagram that shows things fixed in relation to the stars, just as terrestrial maps usually have north at the top. In this kind of astral map, the orbits of asteroids appear chaotic; they are coming at us from all directions! There appears to be no preferred direction in which to look for asteroids; any direction is as good as any other. So why not look only at midnight, and only when the moon is new?

We will describe a different way of looking at asteroid orbits that *does* indicate a preferred way to look. In this analysis, the direction to the stars is not fixed. Instead, we draw our map with the Earth fixed at the time of impact, and show everything relative to that point. The positions of the Earth and the Celestial Object Threatening Earth (COTE) are measured in time from this point. The sun, of course, remains fixed at the center of the diagram.

We know precisely where the Earth will be at any given time, so what we need to do is to work the orbit equations in reverse to see where an asteroid must be now if it is on a collision course. This may seem impossible, since there are many possible paths that an asteroid or comet may follow. However, all of the collision courses that will impact us in the near future are very similar. As we said, the Morrison committee search strategy is to look from planet Earth at about midnight. As will be discussed in detail shortly, the fact of the matter is that objects that are going to hit in the near future spend most of the last few months of their approach on the daylight side of the planet! Thus, they cannot be seen with an Earth-based telescope.

If this is true, then it must follow that most COTEs that come very close to Earth will be discovered only a short time before they approach. This is, in fact, what has been happening. The asteroid that first raised this concern about the impact danger, 1989 FC, was not discovered until *after* it had passed Earth. More recently, 1996 JA1 and 1996 JG were discovered only days before their closest approach to Earth, much too late to have mounted any planetary defense, short of having a whole army of giant, nuclear-tipped missiles standing constantly ready, at a cost of billions of dollars annually.

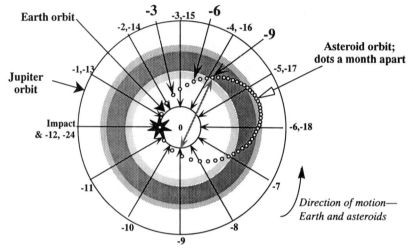

**Figure 6.1: Four-Year Apollo Asteroid Collision Orbit**

Evidently the Morrison committee did not try to discover from what direction the danger might come. If our contention is correct, then we will ultimately need telescopes similar to the Hubble (but much cheaper) somewhere in space, able to look in the general direction of the sun, to see the danger approaching. This concept was specifically ruled out by the Morrison committee, on the basis that it was cheaper to search with ground telescopes.

First we need to assure the reader that all objects in orbit must follow a path that is from the mathematical family known as "conic." As a practical matter, all the paths with which we are concerned are elliptical. Furthermore, all of the asteroid paths which threaten us lie in the same general direction, as we will soon see. We can describe these paths based upon the laws of orbital mechanics and what we have already observed about COTEs. Nearly all of these threats come from farther out in the solar system than the Earth. These may be Apollo class asteroids (named after the first Earth-crossing, and thus, Earth-threatening, asteroid found), or they may be comets. To start, look at figure 6.1. This shows an Apollo asteroid on a path that will collide with Earth.

The inner solid circle is the Earth orbit, and the "O" in the center of the diagram represents the sun. The shaded area indicates the area referred to as the main asteroid belt. The point of collision is on the left side of this circle where the dots representing asteroid positions and the circle of Earth's orbit touch. The dark star shows where Earth and the asteroid impact in their respective orbits around the sun. Each hollow dot shows positions every month before. The arrows which point toward the Earth's orbit indicate the position of the Earth each month before impact. Thus, the arrow labeled "–7" shows where the Earth will be seven months prior to impact.

The reverse direction of these arrows represents the direction the Morrison committee suggested that we look for asteroids. Note that, in this particular example, for a month, and two and maybe even three months before the impact, the midnight direction that the NASA astronomers suggest will allow us to see the asteroid. But this is very late to do something about that impending impact! The gray, two-headed arrow shows the direction in which we should have been looking nine months before impact to see the asteroid, as opposed to where the Morrison committee advised us to look. Note that an Earth-based telescope would be in daylight, facing the sun (the "O" in the diagram), and therefore would be blinded by it. This will not permit an Earth-based telescope to see this asteroid nine months prior to impact.

Consider some other examples from this figure. Note that minus five months is the same position for Earth as minus seventeen. Look at the asteroid position at minus seventeen. You will see that it is is in the midnight sky at this time as well. However, the asteroid is now 2.9 AU away from Earth instead of only 0.4 AU. And it is 3.9 AU from the sun, rather than 1.4. The combination of these factors means that the asteroid is about 400 times dimmer than it is at two months, and therefore, we need a much more powerful telescope to see it. If we look in the dawn sky four months before impact, the asteroid is 100 times brighter than it is at seventeen months, or only four times dimmer than at two months. So by just looking in the right direction, we can gain two months of warning time. More time is needed still, and we will argue later for a space telescope to provide this. You can draw other lines for yourself to see how the situation changes over time.

This example was *not* chosen as a worst case scenario; in fact, as we will show in the next few pages, this is one of the best cases for obtaining an early warning of impending doom. Before we go on, we need to point out that the doomsday asteroid can come from any part of the fixed

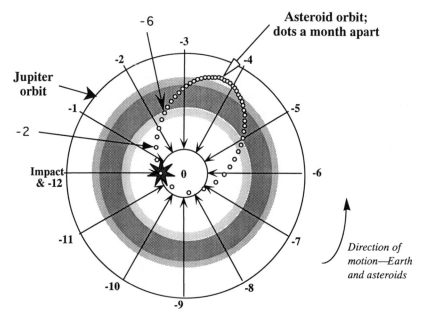

**Figure 6.2: Four-Year Asteroid Orbit; Closer Approach to Sun**

celestial sphere. What we are trying to show here is that its position relative to the Earth and sun will be pretty much the same for any Apollo asteroid. The Earth and the asteroid are going around the sun in the same general direction, and where their orbits intersect will almost always occur on the opposite side of the sun from where the asteroid left the main belt. Because of this, we have some clue about where to look for asteroids that are coming close to us on the trip we both make around the sun. Again, it should be noted that this direction is not fixed with respect to the stars; it is fixed with respect to the direction to the sun.

Figure 6.2 shows another asteroid which has the same period as that in figure 6.1 (it takes the same amount of time to make one trip around the sun), but which has a closer approach to the sun. At one month before impact, the asteroid is almost at Earth local midnight, while for the first example, it was at midnight (where the Morrison committee wants to look for it) about two months before impact. At three months before impact, the asteroid depicted in figure 6.2 is in the dawn sky and cannot be viewed by

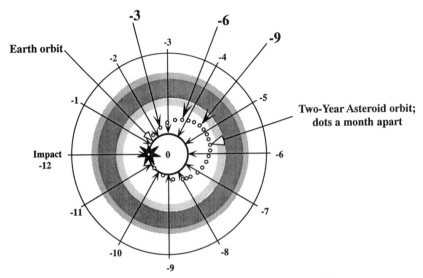

**Figure 6.3: Two-Year Apollo Asteroid
on Collision Path with Earth**

Earth-based telescopes, because the sky is too bright. It is important to note that the midnight view is looking at where this asteroid was about a year ago, and if it is a small asteroid, it was probably too dim to see at that time.

So, this case is *not* better than the first case, as we said before. We therefore need to look at some more cases, in order to discover what may be learned from these diagrams. The shape of these ellipses, showing the path of an asteroid in interplanetary space, depend upon only two parameters, the period of the asteroid, and how close it will come to the sun at its closest approach to that star.

As we have said, most COTEs are of the Apollo class; that is, they come from the main asteroid belt, but cross Earth's orbit. Such an asteroid from the outer edge of the belt will have a period of about four years. This is the period of the orbit just detailed. Asteroids that come from the inner edge of the belt will have a shorter period, but almost none can be expected to have periods less than two years. So, let us examine the path that an asteroid like this must have if it is to impact Earth. Such a case is shown in figure 6.3.

First, note the similarity of this orbit to the four-year asteroid depicted in figure 6.1. The "egg" lays in about the same direction, but it is much smaller, and the dots are much farther apart in figure 6.3. Because the orbit is smaller, the asteroid is moving much faster, so dots a month apart will have larger gaps between them. At one and two months before impact, the asteroid is still in the midnight sky, and at three months before impact, it is about halfway between the minus two- and minus-three-month positions of Earth. This is just what happened in figure 6.1.

In similar fashion, at six months before impact, the asteroid is in the dawn sky, and probably not at all visible from the Earth. The principle difference in this case, then, is that the asteroid is closer to both the sun and the Earth, and hence will be easier to spot. But, to detect approaching trouble early, at or near midnight is *still* not the best direction to look.

## Radar and Lasers

We need to take a minute to explain why we have been talking only about optical telescopes to find COTEs. Why not use radar? Isn't that better? Well, the answer is "Yes, radar would be much better."

Radar, *r*adio *d*etecting *a*nd *r*anging, often used in marine and aircraft navigation, operates by bouncing radio waves off objects. The amount of time it takes for the radio waves to bounce off the object and return to their point of origin indicates the relative distance of the object being tracked as well as its velocity.

Since radar must "illuminate" its target with radio (or light) waves, it must either have extremely powerful transmitters or extremely narrow search beams. Consequently, radar (or ladar—*l*ight *d*etecting *a*nd *r*anging, or optical radar) is not suitable for searching for unknown objects at planetary distances. However, once the direction of the objects is known, it can be used to measure range and velocity. Since lasers have narrower beams than radio waves, they are better suited for this range-finding function.

Unfortunately, the distance at which we need to detect cosmic intruders is *much* greater than we can achieve with even our most powerful radars, even those used to detect approaching long-range missiles.

One of the advantages of radar is that it can determine not only the direction of the threat, but also the range to the threat and the speed with which it is approaching. This advantage is quite obvious when one considers figures 6.1 and 6.2, and examines the distance from Earth these two different asteroids are at midnight, three months prior to impact. Knowing the distance to the object seen by the telescope would be of tremendous assistance in sorting out what is going on.

There is one important possibility not considered by the Morrison committee. They did not examine the possibility of using lasers in a manner similar to radar, i.e., bouncing light waves (rather than radio waves) off objects to determine how far away they are (their range) and their velocity.

Since the start of the Strategic Defense Initiative, or "Star Wars," research a generation ago, the technology of lasers has been greatly advanced, and we believe these improvements should be applied to the problem of defending Earth against COTEs.* One million watts of laser light is now not only quite possible, but actually quite modest when compared to the power required in a laser weapon. Such a laser light, focused into a tight beam (say, one microradian,† another Star Wars achievement), would be far brighter than the sun for tens of millions of kilometers. Such a laser, pointing accurately in the direction determined by the discovering telescope, could tap out a simple pattern of light bursts, alternating the lengths of the continuous release of light by sixty or ninety seconds. (This much time may be needed to allow the signal to be seen.)

The time for light to travel from Earth to the midnight asteroid and return in figure 6.1 has been calculated at 900 seconds and the time in figure 6.2, 2,600 seconds. Thus, an optical telescope fitted with filters to see only the light sent by the laser can readily distinguish between these two objects with even crude timing.‡ With really good timing, this scheme can provide most of the benefits of radar, including both range and range

---

*Initially lasers would be employed to gauge location and speed of an incoming missile. As the lasers become more powerful and sophisticated, they could be used as weapons to destroy missiles at great distances.

†A microradian, one-millionth of a radian, is a very small beam width used to measure plane angle. Prior to the Star Wars research, scientists were not able to focus light into such a narrow, intense beam. For example, if a laser on Earth was pointed toward the moon with a beam this narrow, it would illuminate a spot on the moon only a quarter of a mile wide.

‡Laser light is so pure in color (monochromatic) that there are many more possible laser colors than there are color names. Hence, it is customary when describing laser light to use the wavelength instead of a color name. For example, a blue-green laser might be described as a laser with a wavelength of 4.8 micrometers. Since the sun shines with many colors, looking only at that particular blue-green light and filtering out all other colors allows a telescope to see a light flash that would otherwise be hidden in the glare of reflected sunlight.

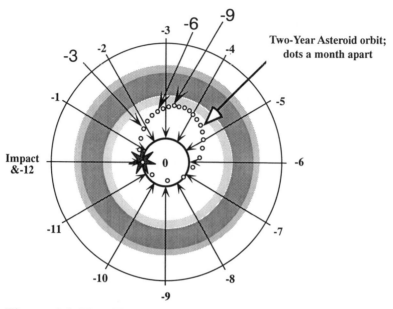

**Figure 6.4: Two-Year Asteroid Orbit Closer Approach to Sun**

rate (i.e., radial velocity). It must be noted that the laser does not need to be located near the observing telescope. Any position within a thousand miles or so would be close enough for this operation to work.* This would permit experimental or military lasers to work with civilian astronomers in locating threatening objects. Laser ranging will locate threatening objects more exactly than telescopic observations alone, and will reduce the search time needed to "re-find" old targets, providing more time to search for new threats.

Now consider the next case, which is depicted in figure 6.4. This is a diagram of a two-year orbit, with a perihelion near that of Venus. Not surprisingly, it is very similar to figure 6.2. In figure 6.4, the asteroid orbit is rotated counter-clockwise, relative to the point of impact, and—if we per-

---

*Since the distance and direction between the transmitting laser and the receiving telescope is accurately known, the delay time caused by this separation can be calculated and a correction applied to the measured transit time of the light pulse.

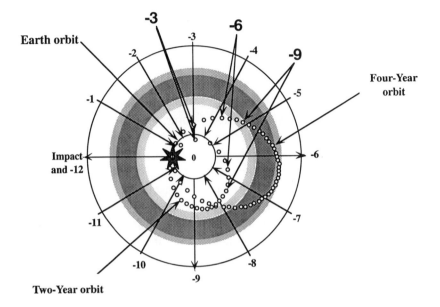

**Figure 6.5: Four-Year and Two-Year Asteroid Orbits That Impact Earth on Second Crossing**

sist in looking only at midnight—we will get very little warning of an impending collision. The positions at each month before impact are quite similar to figure 6.1.

At one and two months before impact, the asteroid is still in the midnight sky, and at three months before impact, it is about halfway between the minus two- and minus three-month positions of Earth. This is just what happened in figure 6.1. In similar fashion, at six months before impact, the asteroid is in the dawn sky, and probably not at all visible from the Earth. The principle difference in this case, then, is that the asteroid is closer to both the sun and the Earth, and hence will be easier to spot.

There is a complication to this description we have given about asteroid positions. If an asteroid, or any COTE, crosses the Earth's orbit, it has two chances to hit the Earth: one on its way toward the sun, and one on its way out. If the impact occurs on the way out, it will change the pictures we have drawn, but it will not change our conclusions about not spending so much time looking at midnight, or the need for space telescopes for more complete celestial coverage. An example of impact occurring as the asteroid travels away from the sun is shown in figure 6.5.

Notice that the ellipses are now tilting down, rather than up. They are, in fact, a mirror image of the previous cases for impact on the trip in toward the sun. As you examine these figures, you see that the four-year orbit is a little bit more favorable for viewing in the midnight sky; at three months before impact, the COTE is almost exactly at midnight of Earth, but six months before it is in the dawn sky, and difficult or impossible to see. And for the six months prior to that, it is lost in the daylight sky again.

The two-year case is much worse. From six to nine months before impact, this COTE moves quickly through the night sky, from sunset at six months, to midnight at seven months, and into dawn at nine months. It is then lost in the sun for another nine months. Thus, based on the examples given, we contend that the case has been made that, most of the time, the asteroids that are going to impact us in the near future are not visible to Earth-based telescopes, because they are on the sunny side of Earth, where telescopes are blinded by the light from the sky.

The Morrison committee, and subsequent studies, have focused mainly on asteroids of one kilometer (0.62 miles) or larger as being the serious threats to Earth. Asteroids this large, astronomers say, may kill a quarter of Earth's population, and probably end civilization as we know it. These "big" asteroids are much easier to see than small ones (which would "only" destroy a single metropolitan area). Hence, it may be true that they will be able to find all of these large asteroids in twenty-five years or so. However, this leaves us vulnerable to large asteroids for decades, and to city-killers for very much longer. *We repeat our call for space-based telescopes to augment the capability of Earth-based ones.*

## The Comet Complexity

We have not yet considered comets. These are more of a problem to find, for various reasons. First, many of them can not be seen until shortly before they impact the Earth. These are the "new" or "long-period" comets discussed earlier. They come through the inner solar system once,

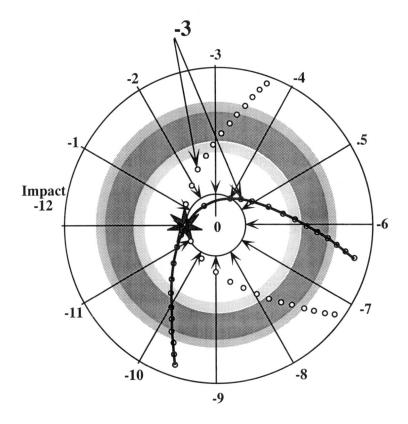

**Figure 6.6: Two Comet Orbits That Impact Earth**

so we have to spot them while they are still far enough away to allow us time to take action. Since they generally do not begin to be visible until they reach the orbit of Jupiter, we have, at best, about a year to find them, determine their orbits, and take action. Second, where they will appear in the sky is less predictable than with asteroids. Figure 6.6 shows the path of two typical comets with near-parabolic orbits, that is, orbits that can just escape from the body being orbited (the term is usually used to refer to long-period comets that are so close to solar escape velocity it is not certain if they originate in the Oort cloud or interstellar space).

The path of open dots shows a comet that just barely crosses Earth's path, and impacts on the first crossing. The solid line shows a comet that approaches the sun as close as Venus, and impacts the Earth on its trip outward from the sun. Again, it should be apparent that looking at the midnight sky will not allow us to see these comets soon enough to give us time to take action to defend Earth. The solid line comet is in the night sky approximately six months before impact, but soon is lost in the day sky a gain. To complicate matters even further, some comets travel in the reverse direction in the solar system, so that these two could be moving in either direction on the paths we have shown. If you study this figure, you will see that this does not improve the situation very much in terms of keeping the comet in the night sky. Thus, we conclude that we will not be safe from the cosmic wanderers until we invest in a space-based warning system.

Where to put such a warning system needs to be the focus of a study by space engineers who are familiar with planning interplanetary missions. There are several factors that will determine the course of that study. The first and most obvious is when the study is undertaken, but other factors include what has been accomplished in finding all of the asteroids (or short-period comets) that threaten Earth and when the new space telescopes, similar to the Hubble—but simpler—can be in place to search the skies. A member of the Space Science Studies Institute, based in England, has recently proposed a very low-cost space telescope, which he dubbed the *Humble* space telescope. The concept is currently being considered by the trustees of a British National Science Centre, and will probably be built in Derby, England.[2] This would cost only millions of dollars, rather than billions, and is the kind of program that we advocate to fill in the gap left by Earth-based telescopes.

If we can get a space telescope in place within a few years to help search for the asteroids that threaten Earth, a good place to put it might be in the same orbit as Earth, but leading Earth by three or four months. That is, the spot in space occupied by the telescope on January 1, for example, will also be occupied by the Earth—but not until April 1 or May 1. This would give the telescope a really good view of the quadrant of the sky from which we expect most of the danger to come. This is demonstrated in figure 6.7

In figure 6.7, the two arrows originating at the Earth's orbit show the direction we need to look in order to find the two four-year asteroids dis-

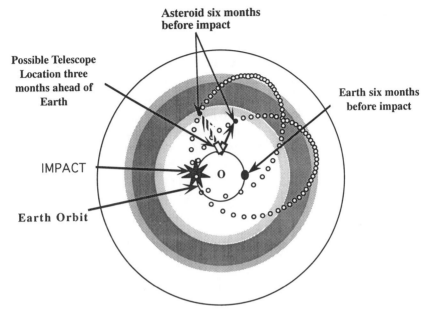

**Figure 6.7: Asteroid Search Telescope Prospective Location**

cussed previously while there is still time to take corrective action, i.e., six months before impact. As has been mentioned, an Earth-based telescope would be blinded by the sun most of the time if it tried to look in these directions.

We need to note that a space telescope preceding the Earth in its orbit by several months has several advantages compared to one located on Earth. The first, which we have noted, is that its view is not obscured by the Earth's atmosphere, so less light is absorbed before it reaches the telescope. Further, sited in this location, it is closer to the approaching asteroids than a terrestrial telescope. And finally, it sees asteroids against a much darker background than is ever possible for an Earth telescope, even on the blackest night. This increased contrast makes it much easier to see dim asteroids (those farther away).

Now, look at figure 6.7. This shows an early warning telescope located in Earth's orbit around the sun, but positioned three months ahead of the Earth. If we ask this telescope to look away from the sun and a lit-

tle to the right (i.e., where the shortest arrow points), it can see an asteroid that will hit Earth six months later. This same asteroid could not be
seen from Earth at midnight (refer to figure 6.1) until two months before
impact. This four-month difference may well be the difference between
success and failure in mounting a mission to deflect this asteroid.

Looking at figure 6.7 again, if the space telescope now looks away
from the sun but a little to the left (shown by a striped arrow originating
at the Earth's orbit), it can see the same asteroid depicted in figure 6.2,
again six months before impact, compared to only one month (if we look
at midnight) before impact for the Earth-based warning system.

In both cases, the distance to the asteroid is greater for the six-month,
space-derived warning than for the Earth-based telescopes. However, this
difference is overcome because the space telescope can see the asteroid
point of light against an absolutely black sky without the dimming effect
of Earth's atmosphere. We argue that the much increased warning time is
well worth the expense of building a space-based telescope.

An early warning space telescope certainly need not cost anywhere
near as much money as the Hubble telescope. That was a research project,
and many high-priced scientists and their graduate students spent many
years charging their time to that project. Here we are discussing a simple
early warning system, which the military knows how to build. The cost
will only be for some engineering, *not* a lot of research, so it can be vastly
cheaper.

If new space telescopes are not to be deployed for ten years or so, and
the Shoemaker plan (which is outlined at the end of this chapter) has been
carried out, then the global emphasis would focus much more upon finding fresh new comets than upon completing the search for asteroids, since
most of the large asteroids will have been found. In this case, the best
thing to do might be to place several telescopes at the inner edge of the
asteroid belt, looking into the quadrant of the sky opposite Earth, from
which most of the intruders will appear. Since these telescopes will move
relative to Earth, several will be required. If cheap access to space, which
will be discussed in chapter 13, has been achieved, then several such telescopes will be quite affordable as compared to the disaster that a comet
impact could cause.

We are forced to conclude that the Morrison committee did not examine the detection problem carefully from the standpoint of determining

what was necessary for providing the earliest possible detection of impending doom, relying instead on the estimate that astronomers would see all potential impact situations years in advance. We certainly hope they are correct, but we are not willing to bet on this assumption.

The balance of this section will demonstrate that the major threats to Earth can be detected in less than twenty-five years, and that, once a threat is found, there are coping techniques available. In fact, once a proper exploration of the asteroids is completed, these planetoids can be used to advance our expansion into the solar system.

## A Follow-up to the Morrison Committee

After comet Shoemaker-Levy 9 crashed into Jupiter, Congress asked NASA for another study of the menace from the skies. The task was assigned to the NASA Solar System Exploration Division in the Office of Space Science. They sponsored a Near-Earth Objects Survey Working Group, headed by Dr. Eugene Shoemaker, now of Lowell Observatory. The group included eleven members from various U.S. agencies and two experts on asteroids from the Russian Academy of Sciences. In June 1995 they published a forty-four-page report, the executive summary of which is reprinted here:

Approximately 2000 near-Earth objects (NEO's) larger than one kilometer diameter revolve around the sun on short period orbits that can occasionally intersect the orbit of the Earth. Only about 7 percent of this estimated population has been discovered. There is about one chance in a thousand that one of these undiscovered objects is destined to collide with the Earth during the lifespan of the average American. Such a collision has the potential of injecting sufficient dark material into the atmosphere to cause a major loss of global crop production and consequent loss of human life.[3]

NASA's charge to this NEO Survey Working group was to develop a program plan to discover, categorize, and catalogue, within ten years (to the extent practicable) the potentially threatening comets and asteroids larger than one kilometer (0.62 miles) in diameter.

Recent advances in the development of Charge Coupled Detectors

(CCDs) for both military (Star Wars) and civilian applications has led to a substantial improvement of capability over what was considered by the Morrison committee. Utilizing these better CCDs on large telescopes could enable rapid progress to be made in a future COTE survey, according to the working group's analysis.

The other very important distinction between the Shoemaker report and the Morrison report is that the former did *not* limit the search area to the midnight sky. The Shoemaker plan will consider the entire sky not obscured by clouds or bright moonlight. This alone, as we have shown earlier, gives it a considerable advantage over the original Spaceguard survey proposed by the Morrison committee.

The recommended program hoped to accomplish the primary objective of discovering 60 percent to 70 percent of short-period NEOs larger than one kilometer within one decade (or by the end of 2006) with funding beginning in fiscal year 1996. That would also put into place the facilities needed to extend the completeness of the survey beyond 90 percent in the following five years. It was also hoped that the Air Force Space Command would allocate an assessment for the defense of the planet within a year or two, as that group had been requested by the Air Force chief of staff to do a mission assessment in that area.

During the Cold War, the Air Force developed a Ground-based Electro-Optical Deep Space Surveillance System, generally called GEODSS, to protect our nation from unfriendly spacecraft. By adding this capability to that of the present Spacewatch system at Kitt Peak, Arizona, the near-term addition of an improved telescope at the same location, and the present facilities at Lowell Observatory, Arizona, the Shoemaker team has cut the time to complete a 70 percent sky survey from almost twenty-five years to perhaps ten. This would still only consider objects of the one kilometer, global catastrophe size, and we believe that the search should be extended to much smaller "city-killers," as soon as possible.

It is instructive to note that astronomy has come a long way in recent years. The first close-approach asteroid, Eros, was discovered in 1898. In 1972, there were a total of thirteen recognized Earth-crossing COTEs. Today you can download a list of over three hundred from the internet.*

---

*http://ccf.arc.nasa.gov/sst/main.html (Three lists are available; Aten asteroids—21 members; Apollo asteroids—169 members; and Amors—131 members.)

Based on the analysis of the discovery statistics of Earth-crossing asteroids (ECAs) to the present and the cratering record of our moon, it is estimated that there are 500,000 to 1.5 million COTEs larger than fifty meters in size—the size of the object believed to have caused the destruction at Tunguska. This is 1,000 times larger than the one-kilometer asteroids that the astronomers are so concerned about. We can expect such a hit about once a century. Tunguska was hit in 1908, so . . .

These smaller threats, along with stray comets, can destroy a city or region. Moreover, if they hit the oceans, which is very likely, they could raise tidal waves that would devastate costal cities thousands of miles from the impact point. We believe that the potential danger from these smaller objects has been seriously underestimated, and that they should receive more attention.

Recently, NASA and the U.S. Air Force have started a cooperative program called NEAT, for Near-Earth Asteroid Tracking, to find more COTEs. It is based upon the GEODSS facilities in Hawaii. This program was announced to the public in April 1996. As of mid-June 1996, 1,715 new objects had been detected. Of these, 252 were tracked accurately enough to be given official designations. Five of these objects—1996 EN, 1996 EO, 1996 FQ3, 1996 FR, 1996 KE—are potential threats to Earth. One new comet has also been officially named: Comet C/1996 E1. All this was accomplished in only a few months of operation, indicating that the sky must be thick with previously unknown objects. At this rate of about one a month, it will be many years before all of the million or so city-killer COTEs can be found.

We applaud this joint venture, which is a partial implementation of the recommendations of the Shoemaker committee. There is more to do, but this is a valuable start. A few million dollars a year is a bargain for planetary insurance. Even more should be done with ground-based telescopes, and planning should begin for one or more low-cost, space-based warning systems. We shall have more to say on this topic in the closing chapter.

**Notes**

1. Duncan Steel, *Rogue Asteroids and Doomsday Comets* (New York: John Wiley & Sons, 1995).

2. *SSI Update* (newsletter of the Space Studies Institute) 22, no. 1 (January/February 1996).

3. *Report of the Near-Earth-Objects Survey Working Group* (Washington, D.C.: NASA, Space Explorations Division, Office of Space Sciences, 1995), p. 1.

# 7

# Islands in the New Ocean

But this is the new ocean, and I believe the United States must sail on
it and be in a position second to none.

—John F. Kennedy, 1962

In the last decade we have learned a great deal about the islands in this
new ocean called space. Not as much as we need to know to settle them,
but enough to pique our curiosity, and make us hunger for more knowledge about this vast, mostly unexplored, territory. In this chapter we will
describe some of the latest discoveries, and show how they fit into an
overall plan for the future.

## Dactyl, Moon over Ida

One of the great surprises of the last few years in the world of minor
planet astronomy was the discovery that asteroid Ida has a moon orbiting
about it. This was wholly unexpected by most astronomers, although in
retrospect it should not have been such a big surprise.

This discovery was made by the scientists of the "imaging team"
(those who study and interpret the information relayed by a space probe)

115

for *Galileo,* the NASA probe that is now orbiting Jupiter. They trained their camera on asteroid 243 Ida on August 28, 1993, as *Galileo* was making its way toward a rendezvous with Jupiter in December 1995.[1] Because the main antenna on the spacecraft was broken, the scientists had to rely upon a tiny back-up antenna to send the data back, which it did, but very slowly. It was February 1994 by the time they were able to finally see the "snapshots" they had taken.

Ida is a fairly small asteroid. Like most of these planetoids for which we know anything about the shape, it looks more like a potato than an orange, with craters instead of eyes. Ida is about thirty-two miles long, giving it a surface area equal to Rhode Island, with enough left over to build the cities of Chicago, Pittsburgh, San Francisco, and Washington, D.C. It may not be a bad place to settle, except that its day is only 4.7 hours long.

Although the picture of the little moon circling Ida took the scientists by surprise, they were quick to realize that this was a great opportunity to discover the density of an asteroid for the first time. If they could discover the period of the little moonlet, Newton's laws of physics would tell them the mass of Ida.* Then, with the volume they had determined from Ida's pictures, they would have the first solid clue for what a main belt asteroid is made of.

Sadly, it did not turn out well for science this time. The little moon's period could not be determined accurately from the few pictures they managed to capture from the wounded spacecraft. The best they were able to do was to set some bounds for the density of Ida, which is believed to be between 2.2 and 2.9 times the density of water. This is in agreement with what was expected, so it did not teach us anything new, except maybe that our previous guesses had been pretty good.

In September 1994 the little moon was officially named Dactyl, from creatures in Greek mythology who lived on Mount Ida in the company of Zeus. It is about a mile long by three-quarters of a mile wide in size, and approaches Ida to within about fifty miles. How far away it gets was not determined, or we would know much more precisely the mass of Ida. It circles its parent not more often than once every Earth day, and probably less.

Ida, at least, is considered to be a member of the Koronis family of

---

*Based upon the relative gravitational attraction each has for the other, the mass of the unknown object (Ida) could be calculated if the period of Dactyl's orbit were known.

asteroids. Richard P. Binzel has studied this family extensively and has concluded that it cannot be more than a billion years old, and may be much less. This suggests it was formed from the break-up of an older, larger asteroid. Hence, it is much younger than the solar system. But both Ida and Dactyl are so heavily peppered with craters that astronomers are led to believe that they must be at least a billion years old, creating a mystery for astronomers to solve.

Dactyl may be another matter, in that it may or may not have originally been a part of Ida. Some spectrographic data suggests that the two objects have somewhat different compositions.[2] Even so, they are thought to be closely related.

Because of Ida's fairly rapid rotation, its elongated shape, and its very low surface gravity, some pretty weird things can happen there. First, take the low gravity. If you drop a rock from eye level, it will take thirty seconds before it hits the surface. According to *Science News*, if you had given that rock some forward motion, it would land in front of you. Sounds normal. But if you threw it forward a little harder, it would double back and land behind you![3] This kind of weird dynamic behavior is being studied to account for the strange location of some boulders on Ida. These house-sized rocks are believed to have been ejected from a crater when Ida was hit by another asteroid. But the boulders are on the *wrong side of the planetoid* for the usual rules of crater formation to apply. The researchers at the University of Arizona in Tucson had to do some complicated calculations to allow for the combination of spin, rotation, and a *very* lumpy gravity field, but they believed they could account for this.

This pair of bodies orbiting together is called a binary system by astronomers. They know of many pairs of binary stars, but the idea of binary asteroids has not been accepted by professional astronomers, even though many amateurs claim to have detected them previously. There are numerous reports from amateur astronomers who say that when they were using a technique known as "star occultation" they detected binary asteroids. This technique depends upon a good prediction that the light from a star will be blocked momentarily by an asteroid passing in front of it. By measuring the length of time the starlight is blocked, the astronomer can deduce how big the asteroid is. But sometimes they see the star blocked twice, and argue that this means twin asteroids. A recent article in the magazine *Astronomy* goes into this subject in considerable detail.[4] The author, Dan Durda, argues, from evidence of dual craters on Earth and the moon, that from 10 percent to 20 percent of COTEs are binary pairs. Some of these are "contact doubles," such as Castalia and Toutatis, both of which are believed to be two asteroids in contact, but some are orbiting pairs, such as Ida and Dactyl are now known to be.

## Gaspra, the First Asteroid Picture

Ida was the second asteroid picture taken by be space probe *Galileo*. The first was Gaspra, and the picture was taken on October 29, 1991. The asteroid happened to lie close to the path of the Jupiter probe, so it was selected as a target for the spacecraft's camera on the long journey. The first image was received on Earth in November 1991. It shows Gaspra to be another potato-shaped asteroid, about twelve miles long and about seven miles wide and high.[5] Judging by the craters on its surface, astronomers think it is *only* 300 to 500 million years old, young for a celestial object. There is, apparently, no moon over Gaspra.

However, Gaspra had its own surprise for astronomers. It apparently has a magnetic field. This was not supposed to happen for bodies as small as Gaspra; such fields were believed to be generated by molten iron inside spinning planets. This cannot be the case for Gaspra, so astronomers have yet another mystery to solve. This magnetic field was discovered when the rest of the data from the 1991 Gaspra encounter were played by *Galileo*'s tape recorder and sent to Earth in November 1992. The magnetometer on *Galileo*'s recorder clearly indicated the presence of a magnetic disturbance in the solar wind. According to Margaret Kivelson, head of magnetic studies for *Galileo,* "[t]he two magnetometer excursions represent *Galileo*'s passage across two wakes in the magnetic field—like a ship crossing the bow wave and stern wave of another craft."[6]

## Toutatis; "Seen" by Radar

Although we have only two optical pictures of asteroids, that is not the only way to make an image of a distant object. Radar can also be used. In recent years, there has been enough interest in asteroids to use the few

Earth radars with sufficient power to make some images. One of these few powerful radars is at the Goldstone Deep Space Communications Complex in California's Mojave Desert. Normally this installation is used for communications with spacecraft in deep space, such as the *Pioneers, Voyager, Galileo,* and others, but when not busy with these duties, this installation's powerful radar can be used for other projects. Steve Ostro from the Jet Propulsion Laboratory in Pasadena, California, used this facility in December 1992, when the asteroid known as Toutatis passed 2.2 million miles from Earth (almost ten times the distance to the moon).[7] Toutatis has been known for several years, and it comes close to Earth about every four years, so astronomers were able to make plans ahead of time to study it carefully with radar. The team obtained radar images of Toutatis which have been published in numerous science magazines.

These pictures indicate that Toutatis is a "contact" binary, meaning it is two fragments either welded together or held in place by a very feeble gravity. The larger body is about 2.5 miles in diameter, and the smaller about 1.5 miles. This is almost big enough to cause another "dinosaur killer" extinction were the asteroid to hit Earth (which is not expected in the next several hundred years). It would certainly be a major calamity for the human race, and many of our fellow species. The existence of Toutatis proves that there are still giant rocks out there than can be doomsday asteroids, and that they come close to us.

There are other asteroids that have been imaged by radar. A smaller one, known as 4769 Castalia, was imaged in 1989. That image showed two half-mile-wide bodies clinging together, but provided little in the way of details. Doubtless more of these COTEs will be examined in years to come as public interest in asteroids mounts.

## Asteroid Color

Determining the color of asteroids may seem like a frivolous activity, but color is quite important in many scientific investigations. Actually, a scientist would not put it in such simple terms; a scientist would refer to "spectroscopic studies," rather than "comparing color chips," but these amount to the same thing. Of course, when scientists refer to spectroscopic studies, they include color bands beyond what the human eye can detect, such as infrared (IR) and ultraviolet (UV) light. These "colors" often provide information that is not revealed by the visible bands of light.

One example of the use of IR and UV bands is the analysis of multicolor images from satellites to determine the health of crops. It has been discovered that crops like wheat that do not have enough moisture (from rain or irrigation) have a slightly different color in the IR rage than healthy crops. These differences cannot be seen by the eye, or in small patches, but a satellite looking at entire fields can alert farmers that their plants need help, and aids forecasters to predict crop yields.

Similar techniques can help astronomers determine the composition of distant cosmic objects. For objects that glow by their own light, such as stars, scientists look for emission lines, which are places where the light is absorbed by the elements and compounds present in the star. For things like planets, moons, and asteroids, scientists look to see how the light is reflected in different colors. By measuring this precisely, they can compare the color of asteroids with earthly minerals and with meteorites to see if they are similar. If the color is close enough, they infer that both the planet and the earthly compounds are of the same composition.

Astronomers have been trying to determine the composition of asteroids through this technique for years. Most of the meteorites that fall to Earth and are recovered are stonelike, and are composed of minerals with names like olivine and pyroxene. Astronomers have been searching for many years to find asteroids of the same color, so they could be more confident of their assertion that meteorites are chips off old asteroids. They have been singularly unsuccessful until recently.[8]

There were two theories regarding why their search had been futile. One theory contends that some form of "space weathering," presumably caused by solar radiation, the solar wind, and dust impacts on asteroid surfaces, had changed the color sufficiently that astronomers could not find a match with the meteorites. The other theory argues that only small, seldom observed asteroids are of the right color.

The latest report seems to verify the latter theory. A team of scientists headed by Richard Binzel focused their telescopes on a main belt asteroid, 3628 Boznemcova, and found a good color match. It is certainly a small asteroid, only about 4.3 miles in diameter. And it is located right where the gravity field of Jupiter stirs things up, so chunks knocked free could readily be deflected into orbits that would bring them to Earth. The scientists note that, "Although the spectrum of asteroid 3628 differs from that of any main-belt . . . its closest similarity is with that of the near-Earth asteroid 1862 Apollo, previously identified as a potential inner solar system source body for ordinary chrondite meteorites."[9] This evidence ties the meteorites more closely to asteroids than comets and is further evidence that it was an asteroid, not a comet, that did in the dinosaurs. This discovery also means that we can use the evidence from meteorites to infer the composition of asteroids.

This view is reinforced by some scientists who are involved with the Near-Earth Asteroid Rendezvous (NEAR) mission to Eros just launched in February 1996 (further details on this mission are given in chapter 8):

> Ordinary chondrites, which make up some 80 percent of the meteorites in museum collections, are primitive assemblages of iron-nickel metal and silicate minerals that have changed little since they formed from the solar nebula some 4.6 billion years ago. Basaltic and stony-iron meteorites, which make up about 10 percent of the collections, bespeak instead active geology in which melting and vulcanism occurred in the first hundred million years of the lives of their source bodies, presumably some of the larger asteroids.
>
> The composition, bulk properties, and provenance of S asteroids are key links in establishing the connection between meteorites and the history of asteroids, and in better quantifying the nature of the impact hazard that the asteroids pose to Earth.
>
> Most asteroids reside in a broad belt between Mars and Jupiter, but a substantial number lie in orbits that bring them close to Earth. These are the "near-Earth asteroids," prime candidates for the origins of meteorites. Dynamicists who study the evolution of asteroid orbits believe that the near-Earth asteroids or NEAs are mostly pieces cast out of the main asteroid belt by the gravity of Jupiter, with some fraction of extinct comets.[10]

## Predicted D-Day: August 14, 2126

In October 1992, a comet, estimated to be six miles in diameter, similar in size to the comet or asteroid that hit Earth 65 million years ago and wiped out the dinosaurs and two-thirds of all life on Earth, was spotted making a pass near us. That comet, named for its two co-discoverers, astronomers Lewis Swift and Horace Tuttle, was predicted to make another, closer pass of our planet with a calculation that it has a one-in-ten thousand chance of hitting the Earth.

The date for this interloper's return to our environs has been calculated to be August 14, 2126 (143 years from now). By that date, humankind should have been able to perfect an antidote to divert such an incoming cosmic object before disaster befalls our offspring. Comet Swift-Tuttle presents a challenge that should wake us up to the potential life-threatening peril of a doomsday rock. The fairly good odds that the comet will hit our planet should provide the incentive necessary to prepare for its return, as well as to prepare to defend ourselves against other asteroids and meteorites that telescopic and radar scanning of outer space can detect in the interim.

We have been warned, and a lengthy period of inaction in the near future will require a much more intense effort to detect and divert any comet or asteroid intruder. If we delay and the incoming object gets too close to us, it could require a giant nuclear space explosion with a blast 100,000 times larger than the first atomic bomb in order to save ourselves (see chapter 16).

Some critics have pooh-poohed this doomsday scare and have accused the astronomers and bomb makers of trying to stir up future business. But the doomsday scientists say the threat is real, and they reiterated the well-known fact that space is swarming with thousands of Earth-crossing asteroids, meteorites, and comets.

Comet Swift-Tuttle got its name during the Civil War when it was first sighted on July 16, 1862, by an American astronomer, Lewis Swift, in upstate New York. It was seen independently three days later by another astronomer, Horace Tuttle, at Harvard University. In September the comet brightened into an object visible to the naked eye, and it later became known as the Great Comet of 1862.

Many present-day astronomers believe that Swift-Tuttle causes the annual Perseid meteor showers, which usually peak in August. The comet's repeated passes through the solar system over the ages, all the while shedding bits of dust and debris, has formed a river of particles along its path. As Earth passes through this river each year, the sky comes alive with flashes of light. The Perseids are the best known and most reliable of the many meteor showers seen from the Earth each year.

Swift-Tuttle's placement in the well-known Mars-Jupiter orbital band was something of a mystery until late in September 1992, when Tsuruhiko Kiuchi, a Japanese amateur astronomer, rediscovered it using a large pair of binoculars. This sighting was one of the most important astronomical finds since the reappearance of Halley's Comet nearly a decade ago. Swift-Tuttle's close approach to Earth on November 7, 1992, meant that the Perseid shower in August 1993 would probably be unusually bright due to the comet's recharging it while on its close passage to the Earth.

But the reappearance turned ominous when, on October 15, 1992, the International Astronomical Union, the world's leading authority in the field, issued its first warning on of a potential collision between the Earth and a large object, with a slight chance that the comet might strike the Earth on its next pass. Dr. Clark Chapman, an astronomer at the Planetary Science Institute at Tucson, Arizona, a private research group, says: "One in 10,000 is not an infinitesimal risk [of being hit by a comet in the twenty-second century]. If the comet turns out to be a serious threat, our great, great grand-children would have to deal with it, not ourselves. . . . One of the problems still to be deduced is that astronomers have no idea at this time as to how much the comet's orbit will be shifted due to the disruptive [cosmic] forces working on the comet's surface, which increase as it nears the sun."[11]

Other astronomers agree with the estimate of a one-in-ten thousand chance of a hit by the year 2126, but Dr. Alan Harris, a planetary scientist at the NASA Jet Propulsion Laboratory in Pasadena, California, noted that the exact extent of the risk could long remain a mystery since forecasts could be rendered obsolete by unrecognized gravitational forces working on the comet in the icy fringes of the solar system, beyond the ken of the Earth's telescopic observatories.

Dr. Johndale Solen, a physicist at the Los Alamos National Laboratory observed that when the comet comes back in the twenty-second century and moves within the plane of our planets, it would be about four years (or 2122) before a potential collision with the Earth could occur. At that point, a nuclear explosion equal to about 10 million tons of high explosives like TNT would be needed to nudge the comet's orbit so it would just miss Earth. Yet to be prudent, he encourages the use of a larger nuclear explosion on the order of 100 million tons as a back-up insurance policy.

When a future comet or asteroid impacts somewhere on the Earth at one hundred times the velocity of a speeding bullet, slamming into terra firma or the ocean while travelling at 20,000 miles per hour, the resulting cataclysm is awesome to contemplate. Great tidal waves could result and vaporized rocks could be hurled into the heated atmosphere as the air turns to a hot pink glow. Buildings and trees would burst into flame while the oxygen and nitrogen in the atmosphere turn into nitric acid. Any form of life that survived the initial impact and was capable of crawling out of a protective cave would be immediately pelted with an acid rain that would sear the lungs and soon thereafter kill most of the few survivors.

But humanity now has it within its grasp to prevent this holocaust from occurring.

## A Recalculation: Doomsday Postponed

Three months after astronomers warned the people of the Earth that Comet Swift-Tuttle might hit us sometime in 2126, some new data was revealed by the Central Bureau of Astronomical Telegrams at the Harvard-Smithsonian Center for Astrophysics in Cambridge, Massachusetts. This new information projected a postponement of the hit by a thousand years or so. Dr. Brian Marsden, the head of the center that issued the original alert from the International Astronomical Union, announced that the new calculations, based on more precise observations, virtually guaranteed that the six-mile mountain of ice and dirt had "virtually no chance of colliding with the Earth," as reported in the December 1992 issue of *Nature*. "We're safe for the next millennium," he said.[12] (Although this latest finding ruled out a comet hit, it did not rule out a possible asteroid hit during that period from another stray near-Earth visitor that may penetrate our orbit when we least expect it.)

The discovery of new asteroids like Ida, Gaspra, and Toutatis can be compared to the "new" volcanic islands discovered in the world's oceans over the past five centuries by our water-bound explorers. As the volcanic islands supplied new information about the Earth, so too will the new asteroids, when they are found, provide clues toward understanding our solar system.

The next chapter will give us a frame of a reference as to how we can best approach a closer understanding of asteroids: first by conducting primitive, unmanned fly-bys; then using asteroid orbiters and manned fly-bys; and ultimately landing on the planetoids themselves.

## Notes

1. J. Kelly Beatty, "Ida and Company," *Sky & Telescope* (January 1995): 20–23.
2. Ron Cowen, "Ida's Moon: Not a Chip Off the Old Block," *Science News* 145 (June 11, 1994): 383.
3. Ron Cowen, "Idiosyncrasies of Ida," *Science News* 145 (April 1, 1995): 207.
4. Dan Durda, "Two by Two They Came," *Astronomy* 23 (January 1995): 32–35.
5. J. Kelly Beatty, "A Picture Perfect Asteroid," *Sky & Telescope* (February 1992): 134–35.
6. "Asteroid Gaspra Surprises Astronomers," *Astronomy* (April 1993): 20–21.
7. Richard Talcott, "Toutatis Seen with Radar," *Astronomy* (April 1993): 36–37.
8. Richard P. Binzel et al., "Discovery of a Main-Belt Asteroid Resembling Ordinary Chrondite Meteorites," *Science* 262 (December 3, 1993).
9. Ibid.
10. Scott L. Murchie, Andrew F. Cheng, and Andrew G. Santo, "Encounter with Eros: The Near-Earth Asteroid Rendezvous Mission," *Lunar & Planetary Information Bulletin* (Spring 1995).
11. Clark Chapman, *International Astronomical Union Report* (October 15, 1992).
12. Brian Marsden, *Nature* (December 1992).

# 8

# Island Hopping in the New Ocean

This deeper understanding and knowledge (of the universe) will bring
the power to predict, to direct and control the forces of nature and our
own destiny.
—Dr. Hugh Dryden, the late first Deputy Director of NASA

It is quite clear from what we have said in the last chapter that we need a
great deal more information about the planetoids before we can either exploit
this vast natural resource or repel one of these cosmic invaders. In short, we
need to do a *lot* more exploring, and we believe that we need to do it quickly!

In this chapter, we will describe some of the plans already in place for
learning more about COTEs, and then tell how and why we think those
plans are not adequate to meet the threat. We need a more vigorous asteroid exploration plan, and we need it now! We will also offer some suggestions that might help speed things up.

## Tools for Asteroid Exploration

When Donald Cox and Dandridge Cole wrote *Islands in Space* more than
thirty years ago, they described six types of unmanned spacecraft that they

126

urged be sent to explore the asteroids. In ascending order of importance, these six main types of future probes were:

1. Asteroid fly-bys to take pictures and otherwise use remote sensors on the asteroid. This was done in an "oh, by the way," fashion when the *Galileo* spacecraft photographed the Gaspra planetoid in October 1991 from 1,000 miles away. *Galileo* was en route to Jupiter, but was programmed to take and relay the pictures back to Earth, marking a dramatic first in the history of spaceflight. This was a fine start, but not the prime mission of *Galileo*, only an afterthought.
2. Planetoid belt fly-throughs (perhaps combined with Jupiter probes);
3. Planetoid orbiters (modeled after other successful probes such as *Lunik III*, a lunar orbiter; the *Viking* orbiter at Mars in 1976; the *Magellan* orbiter at Venus in 1990, and many others);
4. Hard-landing probes (similar to the lunar *Rangers*);
5. Soft-landing probes (similar to the lunar *Surveyors*); and lastly,
6. Planetoid crawlers, a mobile vehicle to move around over the presumably bumpy surfaces, like the one which the *Apollo* astronauts rode on the moon.

It should be noted that in 1964 the timetable projected by Cox and Cole for these various steps was to have the first fly-by missions conducted in 1967, and the soft landers and surface crawlers in operation by 1972, some twenty-five years ago! This schedule was certainly technically feasible, given the pace and momentum of the space program in the early 1960s. The notion that we would abandon all space flight beyond low-Earth orbit by 1972 was utterly incredible. No one could have predicted then that no new space leader would arise to replace Werhner von Braun,* that NASA would become the unwieldy bureaucracy of today, and that the public would become so indifferent to space as to permit our thrust into the solar system to be thwarted.

---

*Werhner von Braun (1912–1977) wa a pioneer experimenter with rockets. He was born in Germany, where he was associated with army rocket research and headed a test facility. He emigrated to the United States in 1945 and became a naturalized citizen ten years later. He designed rockets for the U.S. Army, served as deputy administrator of NASA (1970–1972), and developed the Saturn V rocket used in the *Apollo* flights.

We will now consider these six unmanned types of planetoid-inspecting instruments more carefully.

## Fly-by and Fly-through Missions

As we have noted, the first fly-by mission has been done by *Galileo* as a subset of the mission to Jupiter. Further, there have been a number of missions that have flown through the asteroid belt successfully. The first of these was *Pioneer 10* which was launched on March 2, 1972. James Chestek laid down the mission plan for that journey as the study manager on a NASA-Ames contract.[1] There have been a number of asteroid belt-crossing flights since, thus abating our initial fears of spacecraft being smashed by asteroid dust. We now have accurate measurements of this dust density, and it need not deter us from flights into the asteroid belt.

## An Asteroid Orbiter

The next step in planetoid exploration, if all goes according to plan, should already be well underway. In February 1996, a spacecraft called *NEAR* (Near-Earth Asteroid Rendezvous) was launched from the Kennedy Space Center on a several-year voyage to look at some of our neighbors in space. Its prime target is Eros, the previously discussed asteroid that comes reasonably close to Earth, although it is not an impact threat at this time. On its journey to Eros, *NEAR* will venture into the main asteroid belt in order to set the right conditions for its rendezvous. As it does so, it will conduct a fly-by of the asteroid Mathilde about the end of June 1997. Since its instruments are optimized for asteroid research, that should prove to be a very enlightening time.

In January 1998, *NEAR* will be in the vicinity of Earth, using a "gravity assist" maneuver to adjust its trajectory for the last leg of the trip to Eros. It is supposed to arrive there in January 1999. Once there, it will go into orbit about the little planetoid; it will be the first spacecraft ever to get so close to an asteroid, or to orbit such a small body. It is planned that *NEAR* will spend a year or more making measurements of Eros. It will start its orbit high over Eros, both to get an overall view and to assure mission safety. Then *NEAR* will lower its orbit in stages, until it is only twenty miles above Eros. This will provide very detailed pictures and allow the other science instruments to pinpoint their findings.

Eros is an important target for investigation since it is by far the largest and most important of the near-Earth asteroids. It accounts for over half the volume of all near-Earth asteroids. Eros is a very elongated asteroid, with estimated dimensions of 35 × 15 × 13 kilometers. Earth observations of its color have determined it to be among the stony-type asteroids. However, it is known to be compositionally varied, with opposite sides having slightly different mineralogies.[2]

SOME MAJOR NASA DISAPPOINTMENTS

*NEAR* is the first spacecraft of a new series being developed by NASA under the "Discovery" concept. The *Galileo* probe to Jupiter was a billion-dollar program, and it is going to return only a fraction of the information it was supposed to because its primary antenna froze and failed to deploy in space. The *Mars Observer,* another expensive spacecraft, vanished from the radar screens just as it approached Mars and was about to become useful. (The cause of the disappearance is unknown.) There was so much criticism of the agency for these failures, putting "too many eggs in one basket," that it has developed a new concept, called the Discovery series. Each mission in this series is planned to have a total budget of under $150 million and a development schedule of less than three years. The agency hopes that this will lead to more successful science missions and less criticism if another should fail. Several missions using this concept are now underway, the first of which is *NEAR*.

As a Discovery mission, *NEAR* faced a major challenge of delivering first-rate science on a limited budget and tight schedule. This challenge was met by the design team at Johns Hopkins Applied Physics Laboratory (APL) through clever design tactics. For example, the main antenna is fixed to the body of the spacecraft, so that a *Galileo* failure is impossible. Moreover, the instruments are fixed to the spacecraft body. This further simplifies the design, saves money, and minimizes failure potential. The instruments can be pointed at a fixed locations on the asteroid by maneuvering the spacecraft. They can even track a specific surface feature to build up stereo imagery.

One unusual instrument is a laser range finder. As *NEAR* orbits Eros, it will measure the distance from the planetoid every second, to an accuracy of about twenty feet. This will determine the profile of Eros in great detail, and permit any number of scientific inferences to be drawn about the history and composition of Eros. Note that this project includes the a planetoid orbiter, third phase of asteroid exploration called for in *Islands in Space.*

## Landing Probes: From NEAR to NEARS

Even before the Johns Hopkins University *NEAR* project got off the ground, an added project which would comprise another segment of NASA's new Discovery program was proposed as a follow-up to the main mission. It is called *NEARS*, meaning Near-Earth Asteroid Returned Samples. Its mission objectives are different than *NEAR*, which is aimed at just a close asteroid fly-by. *NEARS* wants to obtain up to six bulk samples of 100 grams (about a quarter of a pound) from an asteroid other than Eros, and then return them to Earth for laboratory analysis. Such an analysis of a sample return would permit a full range of petrological, chemical, age, and isotopic characteristics of what comes safely back to Earth. In a sense, the *NEARS* mission is a "piggy back" on the *NEAR* mission, and is derived from the Hopkins' Applied Physics Lab (APL) development of the *NEAR* spacecraft.

The remote sensing instrument suite on *NEAR* would be replaced by a re-entry capsule to be designed by the Lockheed-Martin company, the same firm that designed the Jupiter probe. This new capsule would contain a pyrotechnic-powered sampling device designed by the APL. The spacecraft would just make a "touch and go" landing on the asteroid. At the instant it touches the asteroid, a "six gun" package would touch the surface, and explosives would drive cans into the asteroid.[3] The cans are then withdrawn, taking with them six samples of the asteroid. The sampling device then retracts into the return capsule, and the *NEARS* spacecraft departs on its return to Earth. As it approaches Earth, the re-entry capsule separates from the rest of the spacecraft, re-enters the atmosphere, and descends the rest of the way by parachute.

This *NEARS* concept was developed by a small team of four members of the APL at Laurel, Maryland, under the guidance of Eugene Shoemaker, the principle investigator for the project. Additional assistance was provided by members of the NASA Lewis Research Center in Ohio to make the *NEARS* mission a "pathfinder for future robotic sample return missions and the logical next step after the pioneer *NEAR* mission,"[4] in the view of the design team.

The *NEARS* idea was supported by the NASA Solar System Exploratory Division, the group that came up with the idea of sending *NEAR* on a four-year round trip to the asteroid 4660 Nereus. The projected launch date for this mission has been set for January 2000, with a return in February 2004. Nereus is completely different from the first three asteroids targeted by the *NEAR* missions. This C-type asteroid a primitive type of minor planet, while Gaspra and Ida, both photographed by *Galileo* during fly-bys, are known to be S-type asteroids, as is Eros, which has been picked for the first *NEAR* mission.

This is rather similar to the *Ranger* series of probes that crashed (what is known as a "hard landing") onto the moon before the first "soft" landers, the *Surveyor* series, were developed. Unlike the *Ranger* probes, which did not survive their landing, *NEARS* would not only survive, but would return an asteroid sample. There may still be some uses for a "hard lander," or penetrator, as it is termed today. These will be described later.

JAPAN MOVES INTO THE ASTEROID HUNT

In October 1995, representatives of the Institute of Space and Astronautical Science (ISAS), the main space agency of Japan, announced in Oslo, Norway, at an International Astronautical Federation (IAF) conference, that their country was developing the first planetary spacecraft designed to automatically land on an asteroid, gather a sample, and return it to Earth for analysis. The project is called *Muses C*. The spacecraft is planned for a launch to the asteroid Nereus in 2002, landing in 2003, and returning to Earth in 2006.

To aid the Japanese program, NASA and Johns Hopkins APL, with the help of the *NEAR* project team, have begun to evaluate whether they could modify the mission plan for *NEAR*. One proposal being considered was to gently "bump" *NEAR* into Eros, much as *NEARS* proposed to do on a later mission. Some project managers felt that contact with Eros would gain valuable experience to aid the *Muses C*. Such a maneuver, however, would be high-risk, since *NEAR* was not designed as a lander.

Space managers at the IAF felt that the proposed *Muses C* mission showed that Japan's maturity in advanced planetary exploration compared favorably to that of the United States, Russia, and Europe. They felt that the time had come to widen the cosmic cooperation to include the Asian nation. Thus, the *Muses C* project marks a turning point in making asteroid detection and deflection a truly global mission involving all of the leading technologically advanced nations—and not just the big three players (i.e., the USA, Russia, and the European Space Agency).

Thus, the first five out of six asteroid missions proposed thirty years ago are all now accomplished or definitely planned. This has not occurred as quickly as hoped, but it happened in just the order proposed.

## A Planetoid Crawler: Does It Still Make Sense?

So many things have happened in the last thirty years that we need to stop and ask if we still need a planetary crawler as proposed by Cox and Cole. Since we have discovered the abundance of COTEs, our first attention will be to them. The main belt asteroids will have to wait for a bit while we assess this newly discovered feature of our solar system. These close-approaching asteroids (or extinct comets, if that is what they are) represent both a promise and a peril. The peril is obvious from what we have said thus far, but the promise may not be quite so apparent just yet.

These asteroids represent the easiest extraterrestrial bodies to get to, in terms of rocket power. They are easier for us to reach (and return from) than even the moon, and are by far easier in every respect than Mars. Further, we know that at least some of the materials we need to colonize space, mainly water, is present on the asteroids. It may be abundant if we land on extinct comets, but even on asteroids it is present in the minerals. So, coupled with our need to explore the asteroids for our own future safety, the lure of knowledge is a "carrot" to lead us to explore these islands in the new ocean.

Now one feature of these islands is that the first ones we want to visit are all very tiny, at least in terms of their surface gravity. A 175-pound person standing on Eros, the target of *NEAR*, would weigh only a quarter of an ounce. It would be very hard to walk without stepping off into space; that is, taking a step would push you off of the planetoid into space on an escape trajectory! Not a good move. In order to remain on the surface, it would be necessary to drive stakes into the planetoid to keep from flying off. Thus, for such an object, a wheeled vehicle does not make sense. What we will want to do is to use very feeble little rockets to move around. We land, and drive in our stakes for a tie-down. When we want to move our spacecraft, very small rocket motors, using only ounces of fuel, will move us to the next place we want to explore. So, a soft lander is the last word in small asteroid transportation; it can land over and over again in hop-scotch fashion with far less weight penalty than a set of wheels. For a large asteroid, such as Ceres, where the mass of a person is measured in tens of ounces, a wheeled roving vehicle may make some sense. However, it will have to move slowly, lest it bounce high into the sky of Ceres and lose traction for minutes at a time.

## More Discovery Missions, then Comets

The first two missions in the Discovery program that we described earlier will be launched in 1996, in February and December respectively. *NEAR* is scheduled to be the first. The second will be the *Mars Pathfinder,* designed to place a small lander and robotic rover on the surface of Mars by July 1997. This mission is intended to accomplish many of the objectives of the *Mars Observer* that failed in 1993. The third mission will be a lunar orbiter which will search for resources on the moon, especially water at the south pole. We applaud these missions, but they are not germane to our subject, so we will not describe them further.

The fourth Discovery mission has recently been announced. It is called *Stardust,* and its purpose is to obtain samples from a comet and return them to Earth for analysis. *Stardust*'s mission is to go into the head of Comet Wild-2 and collect samples of the dust emitted from the nucleus of the comet. For this purpose, *Stardust* will use an unusual material called aerogel, a porous, extremely low-density material that can absorb large amounts of gas or particle matter.

Comet Wild-2 is known as a "fresh comet" because in 1974 its orbit was deflected from much farther out in the solar system by the gravitational attraction of Jupiter. Because of this freshness, *Stardust* should give us a good look at the material in the solar system most similar to that of Oort cloud comets, those which arrive with no warning (and therefore the ones whose composition we need to know before we have to confront them). *Stardust* will approach as close as one hundred kilometers (sixty-two miles) to Wild-2's nucleus on its fly-by in January 2004. The return capsule carrying the dust samples would parachute to Earth, landing on a dry Utah lake bed in January 2006.

*Stardust* will also carry an optical camera that should return cometary images with ten times the clarity of those taken of Halley's Comet by previous space missions, as well as an instrument called a mass spectrometer, which was provided by Germany to assess the light from the comet and perform basic compositional analysis of the samples while in flight.

An even more ambitious plan has been announced by the European Space Agency to send a *Rosetta* spacecraft into orbit around a comet early in the next century. The mission timeline starts with the launch from Kourou, French Guiana, in January 2003, using the European booster *Ariane 5*. The mission makes generous use of planetary billiards (the gravitational pull of the planet) to gain the energy needed to reach Comet P/Wirtanen in August 2011. (As the rocket reaches a planet it can use the gravity to push it toward its next destination.) Two comet landers, named *RoLand* and *Champollion*, are included in the plan. They will land at a relative velocity of about ten miles per hour and will transmit data from the surface of the comet over the next eighty-four hours to the spacecraft for relay back to Earth. The spacecraft will remain with the comet and make observations through perihelion on October 21, 2013.

During this long outbound trip, *Rosetta* plans to fly-by two main-belt asteroids, 3840 Mimistrobell and 2530 Shipka, providing data on them. But note, this is now the year 2010! The science packages for *Rosetta* and its landers are quite complete and will give us very good information about the comet, but not until 2012. We believe it would be better to learn about these asteroids on a much more timely basis.

## A New Millennium Project

As another initiative intended to blunt criticism of its extravagant ways, NASA has devised a program named New Millennium. The idea is to take advantage of the very latest that technology has to offer and build very small robot spacecraft for the exploration of space. According to NASA Administrator Daniel Goldin, "This is a brand new era in spacecraft design. These small, agile spacecraft will be built on a bench, not in a high bay area. They'll be about a tenth of the cost, and about a tenth of the weight of today's spacecraft."[5]

At these prices, NASA will be able to fly many more missions, and do it more often. This capability is very much needed if we are to mount enough missions to the asteroids to understand them well enough to defend ourselves from possible impact, or, alternatively, to use the resources that they contain.

The first of three deep space missions scheduled to be flown by the year 2000 will feature a launch in 1998 of a tiny spacecraft destined for a fly-by of an asteroid and a comet. Spectrum Astro, Inc., of Gilbert, Arizona, will build the 220-pound minicraft under the direction of the Jet Propulsion Laboratory. The details as to which comet and which asteroid will be targeted were still under study as we completed this book.

Several new technologies will be employed on this first New Millennium spacecraft, such as advanced solar arrays, lithium ion spacecraft batteries, and low-mass spacecraft structures. The spacecraft's science instrument payload will include a miniaturized imaging spectrometer that will make chemical maps of the target asteroid and comet.

The spacecraft will also be the first to rely on solar electric propulsion for its main source of thrust, rather than conventional solid or liquid propellant-based systems. This concept has been studied and tested for many years, and is now considered to be reliable enough for use as the main engines. This is important to our future in the asteroid belt, since this technology is most suited for that region of space. It has very low thrust, which is acceptable for use far from planets, but is very stingy in fuel consumption, which is a big plus for these missions.

## Intensify the Search

Although the list of missions just described is a vast improvement over what has been done about the peril from space in the past decade, we think that our nation, and our world, can and should do a great deal more. *NEARS* is a very promising idea, and since it builds directly upon the hardware and the team that is doing *NEAR,* it should be funded promptly by NASA. True, there is another mission to sample an asteroid (*Muses C*), but there are thousands of asteroids, and we must expect to find substantial differences among them. We need a whole series of asteroid/comet landers to develop the information that we need to plan our response when—*not if*—we find a large object on a collision course with our beautiful blue planet.

Along that line, we would also do well to plan a few more asteroid fly-by missions. Now that we know that there are hundreds of asteroids making a close approach to Earth, it would certainly be possible to design a spacecraft with an orbit that could pass close to several in the course of a few years of flight. As we have seen, the European Space Agency will do this for main belt asteroids (which don't even threaten us), and it should be easier to put one into a loiter mode to inspect several COTEs in a short mission.

The mission, although of a fly-by nature, could do more than produce pretty pictures. A small probe could be detached from the main spacecraft and sent on a collision course with the asteroid. Watching it impact with a zoom lens would tell us a lot about how strong the surface of the object is. Is it a dead comet with a crust that is very fragile, or is it a solid rock? The answers to that question will be very important when we have to decide how to deflect a similar object in the future.

A space-borne laser installed on the fly-by spacecraft could be aimed at the passing asteroid and vaporize a surface sample so that spectrometer measurements could tell us, at least, what the surface composition is. That data coupled with in-situ (i.e., on-site) measurement made by a few landers could go far to characterize the threat to our security. It would also go a long way toward answering questions about the resource potential of these bodies.

## Notes

1. J. H. Chestek, "Advanced Pioneer—Synthesis of System Concepts for a Mission to 10 AU," in *Unmanned Exploration of the Solar System,* Advances in the Astronautical Sciences, G. W. Morganthaler and R. G. Morra, eds., vol. 19 (San Diego: American Astronautical Society, 1965), pp. 757–800.

2. Scott L. Murchie, Andrew F. Cheng, and Andrew G. Santo, "Encounter with Eros: The Near-Earth Asteroid Rendezvous Mission," *Lunar & Planetary Information Bulletin* (Spring 1995).

3. E. M. Shoemaker et al., "Conceptual Design of the Near Earth Asteroid Returned Sample (NEARS) Mission," in *Space Manufacturing 10: Pathways to the High Frontier,* Barbara Saughnan, ed. (Washington, D.C.: American Institute of Aeronautics and Astronautics, 1995), pp. 123–29.

4. Ibid., p. 126.

5. Leonard David, "Incredible Shrinking Spacecraft," *Aerospace America* (January 1996): 20–24.

# 9

# Goldmines in the Sky

Gold . . . is the most precious of all commodities, for it constitutes treasure and he who possesses it has all he needs in this world, as also the means of securing souls from purgatory, and restoring them to the enjoyment of paradise.

—Christopher Columbus

The most controversial of all questions asked of the space official by the layman is, "Is it worth the cost?" This question may be followed by others asking what we at home will get out of these explorations and, in particular, whether it will ever be possible to ship space-mined materials back to the Earth economically. To answer the first question, "Is it worth the cost?" in the affirmative, it is important to make a point that will be emphasized later in this book: The major value of spaceflight to the inhabitants of the Earth will be intangible. The most important products sent back to Earth from the planetoids and from other destinations in space will be new knowledge, new sensations, newly captured (photographed or painted) natural beauties, new philosophical insights, and new understanding of the relationships between humanity and the universe. Spaceflight will make a major contribution to our growth toward a more mature, wiser, and happier race.

138

There will also be commercial applications of spaceflight, including the mining of rare materials for shipment back to the Earth. Since the economic practicality of such an operation has been questioned by several scientists, it will be examined more carefully in this chapter. Any speculation about using planetoid materials which takes place before we make manned landings or contact them with instrument probes must be based on some theory about their chemical composition.

The currently accepted theory is based on the assumption that the meteorites which strike the Earth and the planetoids orbiting the sun that we spot through our telescopes are identical in origin and chemical composition. This would appear to be a safe assumption since we observe some relatively large planetoids (compared to the meteorites) passing close by the Earth, and it is only reasonable to assume that smaller ones exist in larger numbers and occasionally hit the Earth.

Actually, some of the meteorites which struck the Earth in the distant past were of the same order of size (about one mile in diameter) as the smaller of the close-approach planetoids.[1]

The only serious question about the asteroid-meteorite theory is the part played by the comets. It may be that some of the meteorites are debris from these fascinating and spectacular objects rather than from the planetoids. In fact, it is generally accepted that meteors (which burn up in the atmosphere before hitting the ground) and micrometeorites (tiny cosmic dust particles) are the residue of comets rather than planetoids. Also, it has been suggested that the large chunks of ice which have been reported to fall from the sky could be ice meteorites derived from comets. To further complicate the issue, it is suspected that many of the close-approach planetoids themselves are decayed comets which have lost all of the volatile material which produces the fiery tail of the normal comet.

Comets normally have much longer periods of revolution around the sun than planetoids. Comet orbits are long ellipses (or in some cases, parabolas) which extend from perihelia near the sun to aphelia in the far outer reaches of the solar system. We need not try to settle the question of which are asteroids and which are extinct comets here, since our major proposal is that these small members of the solar system be the first visited, exploited, and colonized. It appears almost certain that these small objects are similar to the meteorites and it may not really matter whether they are called comets, asteroids, planetoids, or even meteoroids.

In discussing the meteorites, it is important to distinguish between "falls" and "finds." Hundreds of meteorites have been collected which were actually seen to fall and were purposely searched for and located. Many other meteorites have been found whose falls were not observed. There is a very important difference between these "falls" and "finds" in that almost all of the latter are the iron-nickel type while 93 percent of the former are "stones." This is not really too surprising since the "irons" are much more easily recognized as of extraterrestrial origin (or at least as unusual rocks) and have been found in great number by amateurs. The stones, however, are much more ordinary in appearance and would not be recognized as something unusual except by the experienced collector.

One wonders how many other types of meteorites hit the Earth's atmosphere and are destroyed before reaching the ground, and how many reach the ground and are then quickly destroyed or rendered unrecognizable through oxidation and other weathering processes. If ice meteorites are possible, how many strike the Earth per year? It is impossible to say, now. It will be necessary to collect and analyze samples in interplanetary space.

The largest group of the stony meteorites, which represent 86 percent of all falls, are called "chondrites." Chondrites are characterized by the occurrence of small, fused spheres or spheroids of silicates or other minerals in their interior. These tiny "marbles," measuring only about one-tenth of an inch in diameter, are called "chondrules." The stony meteorites which do not contain these little marbles are, logically, called "achondrites."

Since chondrites do not occur as terrestrial rocks, they can easily be identified and represent a distinct class of mineralogical objects. Presumably the majority of the planetoids will exhibit a similar composition. Chondrites can be broken down into further subclasses according to their mineral compositions, the details of which go beyond the scope of this book.[2] However, the members of one subclass are of special interest. These are the carbonaceous chondrites which comprise about 3 percent of the total and are characterized by the complex hydrocarbons which are found in significant quantities in them. Two or three percent of the weight of these meteorites is carbon in the form of hydrocarbons.

The exact composition of asteroids, of course, remains to be determined by our first visits to these bodies. However, under the assumption that the asteroids will be similar in composition to the chondrites, we can see that they will have great value to our future space colonists because of

## Table 9.1: Asteroid Composition

| Element | pounds per ton |
| --- | --- |
| Oxygen | 2,000 |
| Iron | 828 |
| Silicon | 468 |
| Magnesium | 425 |
| Sulphur | 267 |
| Carbon | 136 |
| Calcium | 60 |
| Nickel | 45 |
| Aluminum | 39 |
| Sodium | 31 |
| Nitrogen | 12 |
| Titanium | 2 |
| Platinum | fraction |

Source: Robert Hutchison, *The Search for Our Beginning*, (New York: Oxford University Press, 1983).

the volatile materials, such as water, that they contain. Further, since this composition is very similar to the composition of the sun, we have increased confidence in what we expect to find on these primordial bodies.

The most common element found in chondrites is oxygen. Table 9.1 lists some of the common elements and how many pounds of them we expect to find for each ton of oxygen. We can see that there will be plenty of materials for construction, such as iron, aluminum, and magnesium. We can use the iron (in the form of steel) for permanent structures, and the aluminum and magnesium for spaceships, which must be light. There is plenty of silicon, which could support a vast industry of semiconductor manufacturing. There is enough nitrogen to support our agriculture. Since this material in already is space, we need not haul it up from the Earth (or the moon, or Mars), making it much easier to develop a vigorous, self-supporting, and profitable space industry.

One very important type of carbonaceous chondrite has been found to contain 20 percent water and a second type, 13 percent water. The ordinary chondrite averages about 0.30 percent water and even this low percentage would be a boon to the space traveler. A planetoid made up of 10 to 20 percent water would be a perfect site for an interplanetary "filling station." The ordinary chondrites comprise 75 percent of all the falls and thus would represent the most probable composition of a planetoid.

One other theory of the origin of meteorites and planetoids should be mentioned since it could have a bearing on the chemical composition of the minor planets we select for bases, colonies, and captured space stations. The most probable circumstance is that the planetoids would resemble the ordinary chondrites in composition. However, this conclusion might not follow if the theory of R. A. Hall, an astronomer and medical doctor, is correct. Dr. Hall argues that most of the stony and stony-iron meteorites are "secondary," they have resulted from impact of a large iron planetoid with a still larger planet or satellite.[3] Some objects large enough to have substantial rocky crusts and thus may be sources for stony meteorites include Mercury, Venus, Earth, Mars, the moon, and the satellites of other planets. Iron planetoids striking these planets or satellites could knock off showers of stones which would later fall on the Earth.

This theory might not affect our plans for utilizing planetoid resources if some of the fragments from those collisions were large objects of several hundred feet in diameter that could be captured or otherwise exploited. The place of origin of the object would not necessarily affect the use of its resources. However, it is quite possible that all fragments of such collisions would be small, a few tons or less, and of little interest to our astronaut prospectors. All or most of the larger objects of several hundred feet in diameter and more might be represented by Dr. Hall's "primary" iron meteorites. If this turns out to be the case it might be difficult to locate an asteroid suitable for production of propellant and life-support materials. However, we now have reason to believe that there are many extinct comets in the COTE population, and we can get water and carbon from them if need be.

That elements exist in COTEs which might be economically returned to Earth is admittedly the most speculative area and probably of rather secondary importance to the space effort of the next few decades. It is, nevertheless, a matter of great interest to many people. Several writers have proposed that in the future it might be possible to mine valuable minerals on the planetoids and ship them back to Earth to ease the anticipated shortages of critical materials. Others, including some leading space "experts," scorn such schemes as unrealistic because of the prohibitively high cost of space transportation. Given the present space shuttle costs, these concerns are quite real. But, as we will explain shortly, these costs can be dramatically reduced by replacing the shuttle with a good space transportation system. Because previous discussions of the return of asteroid materials have almost invariably been qualitative (and therefore inconclusive), it seems worthwhile to introduce a few numbers.

A flight from the close-approach planetoid Eros to the Earth would require a velocity increment of approximately 1,400 miles per hour, assuming correct departure time and a minimum energy trajectory are chosen. The vehicle will arrive at Earth with a high relative velocity which can be removed through atmospheric braking (the friction experienced when an object in space enters the atmosphere). Thus, a one-way, planetoid-to-Earth vehicle need only have a velocity change capability of 1,400 mph, which is easy to accomplish with a relatively small rocket. (Compare this with the rocket velocity of 18,000 miles per hour which is necessary to send an object from Earth into a low orbit; or 5,500 miles per hour to send an object from the moon to Earth.)

A hydrogen-oxygen-powered rocket with a mass of only 200 thousand pounds (not including payload) could push a giant payload of one million pounds from Eros to a trajectory intersecting the Earth's orbit. Although the plane of this transfer trajectory would be inclined about 11 degrees to the plane of the Earth's orbit, this would not affect the initial velocity requirement. It would affect the arrival velocity, however, but it has been assumed that this would be removed in the Earth's atmosphere.

What would it cost to carry one million pounds of payload from Eros to Earth? If we assume conservatively that 200 thousand pounds of propellants are consumed, then the fuel cost per pound of payload will be only one-fifth of the cost per pound of fuel. Thus if we assumed a cost of one dollar per pound of fuel, the specific payload cost would be only twenty cents per pound! Something should also be added for the cost of the vehicle hardware which would weigh about 20,000 pounds. At $100 per pound, this would add another $2 million dollars, or two dollars per pound of payload. Thus the direct operating cost would be only $2.20 per pound of payload! This cost could be raised by a large factor and it would still be economically feasible to return high-cost raw materials (such as platinum) to the Earth from the minor planets.

Of course, we can't say now what it would cost to produce propellants on Eros. Currently it is possible to produce hydrogen and oxygen on the Earth at average costs of twenty-five cents per pound for six-to-one mixture ratios, which is the ratio that vehicles such as the shuttle use. Whether or not one can assume such costs for Eros depends on the level of technical civilization that is attained in our future planetoid colony. However, if we find water, either as ice or in carbonaceous chondrites, all we need is a small amount of machinery and sunlight. It is certainly possible to imagine a small water-to-propellant plant being shipped to Eros, or other planetoid, where it can supply fuel for many return trips.

In any case, it is certainly unreasonable for "experts" to base the cost of transporting freight back to Earth on the present very high price we would have to pay to carry payloads to Eros. The tonnage of freight returned will greatly exceed the tonnage of supplies carried to the mining site, since the required velocity change for the return trip is only a twentieth, 30,000 miles per hour, required for the the Earth-to-planetoid flight. While the economic feasibility of planetoid mining cannot be proved at this time, it should not be dismissed as impossible. After all, study of the percentage composition of meteorites indicates that Eros should contain about 250 trillion dollars' worth of platinum group metals. Perhaps a little constructive thought on planetoid mining would be worthwhile even at this early date![4]

Willy Ley (a German science-fiction writer who coauthored several books with Werhner von Braun) and Russian scientist F. Pokrovsky have both suggested that we could make a profitable business of mining the planetoids for platinum and other noble metals. Space prospecting on the minor planets for rare metals can become—in time—a greater incentive for an expanded space program than the Gold Rush. (Recall that the 49ers provided a great deal of motivation for the pushing of our Western frontier to the Pacific.)

It is worth noting that we are reasonably sure of the presence of platinum in the planetoids. Remember, it was a platinum group element, iridium, that gave us the clue about the dinosaur extinctions. Platinum is a most valuable industrial material. Many of its uses are not practical at present because its scarcity limits it to jewelry and a few other applications (such as the catalytic convertors on automobiles) where its use is so essential that we pay the high price for its benefits. If it were available in much larger quantities, many benefits would be derived from it. So, we need not worry about depressing the price too far by import, since the demand is much greater than any supply we expect from the planetoids anytime soon.

Fortunately, we do not have to wait for the first astronaut-geologist to fly out to one of the small planetoids and bring back the evidence as James Marshall had to over 145 years ago when he panned the first flecks of gold dust at Sutter's Mill in California. Our astrogeologists and space miners of the 1970s and beyond will have the direct evidence from our space probes that rare metals more valuable than gold exist on the planetoids.

An important future economic asset of planetoid mining, for space planning purposes, is the assumption that transportation costs for the return trip to Earth will be comparatively low. While the costs of hauling small payloads to the planetoids might be as high as 100–200 dollars a pound, the freight costs for the return trip might be as cheap as one to ten dollars per pound or less, as we have seen in the example of Eros. Because of the "free" ride resulting from the absence of strong gravity fields on the voyage back, much heavier loads can be hauled with less energy than was consumed on the outward-bound trip, and transportation costs will be correspondingly reduced.

As attractive as the economic benefits of planetoid mining may appear to those geologists presently concerned with the rapid exhaustion of our reserve supplies of certain mineral resources here on Earth, such mining operations may have even more importance as a source of raw materials for supporting space exploration and the colonization of our solar system. The use of this ready supply of space minerals could accelerate the exploitation of space and provide a potentially greater long-range dividend to the stockholders of an "asteroid mining corporation" than the immediate short-range, quick profits that might be made by transporting the valuable cargoes back to Earth for processing. If planetoid mining is for the economic benefit of the inhabitants of the Earth, then it offers some interesting possibilities. If it is instead used for some still undefined greater purpose, then perhaps the resources of the planetoids should be used primarily for these greater purposes rather than for raising still higher the standard of material affluence of the advanced technological societies of the Earth.

How we eventually make use of these astro-mining opportunities will help determine our space goals for the twenty-first century. It must be remembered that the major payoff of the discovery of gold in California in 1849 was not more gold, but the settling of what has become one of the largest and most economically productive states of the union.

## Notes

1. R. S. Dietz, "Astroblemes," *Scientific American* (August 1961).
2. See Gerard P. and Barbara M. Middlehurst, *The Moon, Meteorites, and Comets* (Chicago: University of Chicago Press, 1963).
3. Ralph A. Hall, "Secondary Meteorites," *Analog Magazine* (January/February 1964).
4. E. I. Krinov and I. Vidziunas, in *Principles of Meteoritics* (International Series of Monographs on Earth Sciences Series, vol. 7; Tarrytown, N.Y.: Pergamon Press, 1960) have compiled data obtained by several scientists in studies of a large number of iron and stony meteorites. Their figures show that gold and the platinum-group metals average fifty to sixty parts per million by weight. Percentage compositions do not differ widely between the two meteorite groups. Thus, a planetoid weighing a million million pounds (or one about three thousand feet in diameter) would contain 50–60 million pounds of platinum-group metals. At an average (conservative) figure of $1,000 per pound, the Earth value of these metals would total $50 billion. Eros, at roughly seventeen times the average diameter, and five thousand times the mass of this simple planetoid, would contain five thousand times the amount of noble metals.

# 10

# How to Deflect an Asteroid

We're safe for the next millennium.
— Dr. Brian Marsden, Harvard-Smithsonian
Center of Astrophysics

What happens when we see an asteroid heading our way? Are we doomed, or is there some way that we can prevent it from hitting us? The answer to that question is the old familiar, "it depends." It depends upon many things, but most of all it depends upon when we discover it.

If we discover it in 1996, and if it is going to hit us in a few months, the very best we can hope to do is to determine where it is going to hit the Earth, and then evacuate the region that will be affected by the impact. As we pointed out in chapter 5, an impact in the ocean may be the very worst possibility, since that would raise an ocean tidal wave that could cause massive destruction on several continents. But, if we were to see such a hit coming, all we could do in the immediate future is to get away from the impact point and hope for the best. This is not a very satisfactory answer, and one purpose of this book is to point out that we can, and we believe, *should*, make some preparations so that we have a better answer than "hope for the best."

If we, as citizens of planet Earth, have installed an early warning system for approaching asteroids and comets as recommended in the last chapter, we will know one is coming well before it poses a danger; we can determine roughly when it will arrive and if it is an asteroid or a comet. That may make considerable difference in the manner in which we deal with the threat. Another variable which will affect how we respond to the threat will be the size of the object. If we have been searching for wayward asteroids for a decade or two, and then discover one approaching us, we may be certain it is relatively small and we can be guided accordingly. So, to deal with this question of what to do, let us consider several cases that cover the range of possibilities we may have to consider.

First, we need to consider how to deal with an approaching asteroid. The answer turns out to be "give it a shove," such that it will miss the Earth. Before we describe how that might be done, let us see why that is generally accepted as being the best way to deal with a COTE. What we would most like to do is to vaporize the object so that it cannot penetrate the atmosphere enough to bother us at all. This will be practical only for very small objects, however. Suppose, for instance, that we were to detonate a twenty-megaton bomb at a COTE coming towards us. (Note, it is generally believed that a twenty-megaton bomb was the largest hydrogen bomb that the United States kept in stock during the Cold War.) The energy in a twenty-megaton bomb is about 23 million kilowatt hours. This is quite sufficient to vaporize a small asteroid, of about ten meters in diameter. However, a ten-meter asteroid will cause us no problem; it will be smashed by the atmosphere and will cause no damage to the Earth or its inhabitants. By the time we get to a one hundred-meter asteroid (the ones that the detection committee said were "too small to bother with") it would take hundreds of our largest bombs to vaporize the object. And for an object even a kilometer in diameter, the number of bombs required to vaporize it exceeds by a large factor the number of hydrogen bombs ever made! Clearly, vaporizing these threats is not the answer.

The next option that presents itself is to smash them into pieces small enough that they will not be able to penetrate the atmosphere and cause us any harm. For stony asteroids, it is generally agreed that objects smaller than some minimum size will not cause us any harm, at least if they hit singly. It is estimated that the Tunguska object in 1908 was approximately 60 to 100 meters wide, and it caused a lot of damage. Suppose, then, that we say we want to be sure that no object bigger than twenty meters escapes our effort to fragment an asteroid. The impact of a twenty-meter object would be the equivalent of a *three hundred-plus kiloton* atomic bomb exploding at high altitude. This would cause a very noticeable bright flash in the sky, but would almost certainly not kill anybody. But, if we fracture a hundred-meter asteroid into chunks no bigger than twenty meters, we would get at least one hundred and twenty-five smaller flashes. The whole sky would light up. This may cause no more harm than a lot of sleeplessness and worry, but we can't even be sure of that. But if we break up an asteroid one kilometer in diameter, we have two problems. The first is to be sure that no chunks larger than our twenty-meter size escape to rain down upon us. The second is that, even if we get the asteroid broken up that well, there will be over 100,000 such bombs exploding over our heads. This may very well set fire to everything in sight, deplete the ozone layer in a really big way, or cause other mischief that we have not yet even imagined. This does not look like a good plan either. So we return to our conclusion that we need to somehow shove the asteroid away from a collision course with Earth.

In principle, to shove an asteroid onto a different trajectory is a simple matter. We need to use a jet propulsion device to either push it sideways, or make it go faster or slower so that it will arrive at a point where it crosses Earth's orbit before or after Earth gets there. If time is limited, as it may be if we discover very late that an asteroid is about to hit us, our best option will be to shove it sideways. A rule of thumb in interplanetary navigation is that the amount you can steer your probe toward or away from your destination is the product of the mid-course correction velocity you apply multiplied by the time remaining before you arrive at your destination. Thus, if our Mars probe is about to miss our aim point at Mars by a million meters (one thousand kilometers) and if it still has a million seconds to travel (about eleven and a half days), we can simply deflect the course of our probe by one meter per second to correct the aim. This example is quite accurate, since the numbers are "small." In particular, the fraction of the orbit over which the correction acts is a very small part of the total. Also, the correction is small compared to the current velocity of the probe. This same procedure may be used to calculate the work needed to move a threatening object onto a safe path, provided that the limitations of this approximation are carefully observed. It is important to note, however, that is approximation only works for a small deflection applied shortly before impact. We have seen studies by well-known scientists that did not respect this limitation.

When this procedure is applied to COTEs, we quickly learn the value of warning time. Suppose that we see a one-hundred-meter object coming at us, and we have only two months warning. The first problem is that it will be exceedingly difficult to get an interceptor to the threat in time. To get to the asteroid in time, the interceptor must be placed on a high-energy interplanetary orbit. This cannot be done with existing bomb-carrying military rockets; they are designed to simply go to another continent and cannot achieve such an orbit.

After we solve that problem and we have an interceptor on the right trajectory, how much "shove" must it give to the asteroid? Well, using our simple rule above, we see that if the interceptor arrives two weeks before the asteroid is going to hit us, we have about 1.2 million seconds for our "course correction" to work. The effect of the Earth's gravity field will pull the asteroid toward us, so we need to move it aside more than one Earth radius. For an asteroid, this effect means that we need to move the asteroid about twice the Earth's radius, say 12,000 kilometers. So we find that we need to push the asteroid sideways with a velocity of "only" ten meters per second, or about twenty-two miles per hour. No big deal. But wait! The asteroid has a mass of perhaps 1,500 million tons. To move that by ten meters per second would take all of the space shuttle engines firing for 5.9 days at full throttle. This would take six and a half million tons of propellant. We have just described a *very large* interceptor, one that is well beyond the technological capabilities of the near future.

Since we cannot carry enough rocket propellant to deflect this asteroid, we must consider using a nuclear bomb, since it packs much more energy into a smaller package than chemicals do. This concept was first suggested by Dan Cole in *Islands in Space* many years ago, but the idea was not taken seriously until quite recently. This is the principle means for deflecting asteroids that was proposed by the 1992 Los Alamos conference on asteroid deflection.

Before we show how a nuclear bomb can deflect an asteroid, we need to explain something about nuclear bombs in space. On Earth, any bomb, nuclear or chemical, heats the air to form a powerful blast wave. This causes much of the damage, and is what people imagine when they think about a bomb. In space, since there is no atmosphere, there is almost no

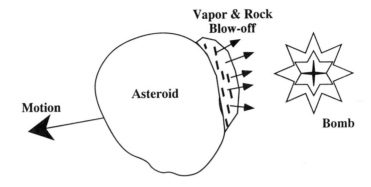

**Figure 10.1: Asteroid Deflection by Nuclear Bomb**

blast effect from the bomb. Only the vaporized bomb material can produce a blast effect, and this is minuscule. Almost all of the energy of a
nuclear explosion in space takes the form of radiation (X-rays, gamma
rays, and so on). This radiation will go in all directions from the bomb, at
the speed of light, until it hits a solid object. In deep space, nearly all of
this energy will leave the solar system, harming nothing. So how can that
move and asteroid?

Figure 10.1 shows a bomb exploding in space close to, but not on, a
COTE. The radiation from the bomb penetrates the asteroid and super-
heats it so that bits of it fly off, either as vapor or fragments. This provides
a jet reaction (like a crude rocket engine) which moves the asteroid the
other way. For a small asteroid, a single bomb will probably provide
enough deflection. If the asteroid is large, it may take several nuclear
blasts to provide enough deflection, depending upon how much time is
available for the deflection maneuver to operate.

Note that most of the radioactive residue of the bomb is carried away
by the fragments that were blown off (and will miss the Earth) so that the
asteroid will still useful to us if we later decide to colonize it. So this atom
bomb deflection technique does not harm Earth or make the asteroid too
"hot" for human use.

Consider next the case of discovering, in the next few years, a *small*
asteroid that will give us only a short time to react, say several months.
This is a very realistic possibility. One extreme means of preparing for

this event would be to build a fleet of rocket interceptors to stand by, with huge nuclear bombs on board, ready to go forth and blow the intruder into very small pieces.

If we were to try this, we have the problem of assuring that all of the asteroid fragments are too small to hurt Earth when they hit, and we are not sure we can do this. There would also be some concern about radioactive residue from the bomb remaining on these fragments, which may fall into Earth's atmosphere. This radioactive residue will be very small, but some people will worry about it.

Such a fragmentation mission would require a huge arsenal of rockets armed with very large nuclear weapons. It would be the Cold War all over again, except that it could be an international effort aimed at the sky and not at another group of humans. This effort would be quite costly in terms of money and labor. Further, there is always the possibility that the people who controlled such power might choose to use it for their own purposes, such as conquering the planet. There are at least some people who believe that the risk to our survival in having a large nuclear-armed rocket force on standby is much greater than the risk to our survival from a celestial object hitting us. In view of this sentiment, we have cast about for an alternative that would deal with this situation but which would *not* pose such a great risk to the peace of the planet.

First, we have to ask ourselves what launch systems would be capable of launching a nuclear warhead and hitting an incoming rock. This question was within the responsibility of the Spaceguard interception conference, but it was not really addressed, at least not in the report that was published in 1993. It was apparently assumed by those assembled that a nuclear warhead could be delivered to an approaching asteroid using an existing ballistic missile. This is simply not so. Most of the world's ballistic missiles could not put more than a fraction of their nuclear warheads into any Earth orbit, let alone into a trajectory that would escape the Earth's gravity field and get into interplanetary space. This point appears to have escaped most of the participants of the deflection conference in Los Alamos in 1992.

Well, if the world supply of intercontinental missiles will not suffice to attack an approaching asteroid, do we have *any* rockets that are capable of this? The answer to this is simply, yes, we do. The rockets that launch our communications satellites into a stationary orbit above the equator have a much greater potential for getting a payload to the approaching asteroid than any of our military rockets.

Now it should be noted that we would have to use these satellite launchers in a very different way than that for which they were conceived or designed. Usually such launchers put a satellite onto a trajectory that will take it to a greater geostationary altitude. However, the satellite at that point is moving too slowly to stay at the required altitude. To prevent it from falling back to Earth, a rocket built into the satellite fires to increase the velocity enough to attain a stable orbit. This operation is routine, and the velocity requirements at each step in the process are well known. *If,* instead of waiting until the rocket is at high altitude before the satellite motor is fired, that motor were fired as the satellite was just leaving low Earth orbit, the satellite would then be on an escape trajectory, bound for interplanetary space.

There are several communications satellite launches each year, generally more often than once a month from one country or another. If some international Spaceguard Defense organization had a few nuclear bombs of the right sort fitted into a satellite body that could be substituted for a communications satellite, then the planet would have a few chances to intercept an incoming asteroid (or comet) before it could impact the Earth and cause a catastrophe. Further, since the bombs are few, are separated from any launchers, and are under international jurisdiction, the potential for misuse is dramatically reduced. Moreover, this program would be relatively cheap. The right bombs already exist, so all that is needed is to build or modify a carrier vehicle to provide the support and guidance to carry this bomb to its destination when needed. A military type organization is needed to protect the bombs, and prepare them for use when ordered to do so by Spaceguard Command. This program would constitute a minimum-expense type of response to the problem of an unexpected asteroid impact.

As time goes by and most of the COTEs are detected, we would have greater confidence that these small atom bombs would be able to deal with any asteroids coming our way, since we would know about all of the big ones, and only small ones would have evaded our detection. Also, as time goes by, we will be able to predict many years in advance where the dangerous asteroids are, and when they will become a threat.

Consider now the case where we know that a particular asteroid will be a danger to us in twenty years or so. As soon as we discover this, we will want to launch a mission to survey the threat up close, to find out exactly what it is, so we will know how best to deal with it. As we discussed in chapter 4, the body in question may be a solid rock, a lump of iron, or an extinct comet, very fragile, but perhaps containing a large store of water or other kinds of ice deep inside. We will deal with these alternatives in different ways. The methods of doing such a survey mission are described in chapters 9 and 14.

Consider the case where we discover that the asteroid is basically a lump of iron. This means that it is massive and very strong. It can probably be most effectively moved by detonating a series of atom bombs on or near its surface. Depending upon the state of our space civilization at the time, we might prefer to conduct this blasting operation with people present to supervise it. If this is the case, we can prepare a cavity where we want the bombs to explode, and to a large degree control the effect. This will reduce the number of nuclear bombs required to complete the operation. Such a manned operation may also give us sufficient control of the asteroid orbit to move it to where it will be convenient for future mining operations, and provide a source of materials for our space colonies.

If this discovery were made only a few years from now, but still with plenty of time to take action, we might prefer to do all of the work with unmanned spacecraft. These could be dispatched to the asteroid on a schedule, following the survey, to explode their bombs in a similar fashion, but with less control. In this case, we will be more concerned about moving the asteroid far enough away so that it will cause us no problems for many years into the future, than about "saving it" as a source of future iron. We will want to use enough bombs to be sure that our mission is accomplished, regardless of the expense.

Suppose, instead, that the COTE is a stony asteroid. Such a rock *may* be too fragile to move with a series of atom bomb blasts. Instead, it may simply be broken into pieces, all of which would hit the Earth, producing more mischief than one big rock. If we have long enough to prepare, we still have a number of options for dealing with this situation. Basically, we will want to use a jet propulsion device such as a rocket engine to move it. There are at least a couple of techniques that can be used in this case. If the composition is similar to the carbonaceous chrondite meteorites found on Earth, there may be sufficient water and other volatile material content to give us a simple solution. We can use the water in a steam rocket to move the asteroid. This technique is really more suited to comets, either extinct or new, so we will discuss it further in the next chapter. We will return to the subject of nuclear steam rockets in a later chapter, when we discuss the possible colonization of the asteroids. For the moment, we will close the subject by saying that this a very promising technique for making small deflections, ones that would save us as long as we had years of warning. However, this approach will be of little help if the warning time is only a year or less.

That leaves us with the case of an asteroid where the only materials that can be used in a jet propulsion device are rock. That will take a little more effort, but we can make a perfectly satisfactory jet propulsion system just by throwing rocks. By throwing in rocks one direction, you create a propulsive force in the opposite direction. This is Newton's Law of Motion,* and every jet engine works upon that principle. In fact, this rock-throwing suggestion for moving asteroids was first made by Cox and Cole in *Islands in Space.*

In that book, the only description provided was of a "linear electric motor." The means of operation was not further specified. It must be noted that since that book was written, a great deal has been done to advance the use of linear motors or electric catapults, which amount to the same thing. There are linear motors that depend upon a magnetic armature to interact with the moving magnetic field produced by coils. Two types of such motors exist, synchronous and inductive. Since they require a magnetic armature, they would be just the ticket for hurling lumps of iron. However, we have postulated a stone asteroid, so that concept will not work directly. However, there are two other forms of electric catapults that will work.

The first concept was proposed by Professor Gerald O'Neil and Dr. Henry Kolm in the book *High Frontier.* This concept is usually referred to as a "mass driver," and was proposed to catapult materials from the moon to a gathering point in space for use in building giant space colonies, or solar power satellites, or whatever large structures are needed in space. The plan basically calls for a conveyor belt which uses magnetic buckets to traverse a series of electromagnetic coils which drive these

*This law states that for every action there is an equal and opposite reaction.

buckets at a high velocity, and then return them to the start to catapult another load. Some people refer to this concept as a "coil gun" to distinguish it from another electrical catapult design usually called a "rail gun."

The rail gun concept received considerable study and development as part of the Star Wars (Strategic Defense Initiative) effort to protect the United States against attacking ballistic missiles. The rail gun is much simpler than the coil gun, since it needs only two parallel rails carrying a very high electrical current for its operation. The electricity travels from one rail, across a gap, and back along the other rail to produce an accelerating force. It has the further virtue that it can propel nonmagnetic and even nonconducting objects. In fact, most of the development work done on rail guns for Star Wars was done using teflon plastic pellets as the projectiles. There are some remaining development problems, but the concept has been well established, and it seems to be only a matter of engineering to build a workable model that could be used as a jet thruster for asteroids, using the asteroid itself for "working fluid," i.e., as the material ejected from a jet propulsion device to cause a reaction force. It would have to be supplied with electrical energy from an outside source, whether solar arrays or, more probably, a large nuclear-electric generator.

There is another type of electrically driven device that could be used to throw rocks fast enough, and in great enough quantity, to move an asteroid. This is a sling-shot. Imagine a hollow tube revolving rapidly about a pivot point. If we drop rocks into the tube at the pivot point, they will slide along the tube and out the end at high speed. This is not very elegant, but it is a simple and effective means of converting electrical energy and rocks into a propulsive thrust. This has never been proposed as a means of propelling space vehicles, since the amount of material that must be thrown away would be quite large in comparison to the velocity change achieved by the vehicle. Since we have lots of rock (the entire asteroid) which we someplace else anyway, this may be a very useful concept, however. The machinery is very simple, low cost, and lightweight. It would be rather easy to manufacture quickly on the asteroid to be moved or at lunar Spaceguard headquarters, to be transported to any asteroid that threatened Earth. Again, these electric propulsion techniques will be useful only if we have years for them to operate.

The next propulsion concept that we want to discuss is the use of explosives, brought to the asteroid from the Earth, or from a moon base, or from Mars, or wherever. This will work, but the energy content of high explosives is modest in comparison with most rocket propellants. The energy content of a ton of TNT is about 1,300 kilowatt hours. The energy content of a ton of hydrogen and oxygen mixture is about four times greater. Now, it is almost certain that we can find the oxygen we need on the asteroid, bound into chemical combinations in the rocks. Since we need eight pounds of oxygen for every pound of hydrogen, if we can liberate the oxygen with a small chemical plant carried to the asteroid, we can reduce the mass of materials that we must carry to the asteroid by a factor of thirty-four compared to taking TNT to the asteroid.

This means that there is a chance to move small asteroids by using blasting techniques. A small cavity would be dug into the rocky asteroid, and then a thin plastic bag of gas would be covered by tons of rock. When the gas mixture is ignited, the resulting explosion would hurl the rock away at some modest velocity, resulting in a push to the asteroid. If this were done at the perigee (i.e., when the asteroid is nearest the object it is orbiting) of the asteroid orbit, several years (three or four) before the impact with Earth, then only a centimeter per second (or so) of velocity change is needed. A few hundred such blasts could change the asteroid velocity that much. If we had a two hundred-meter asteroid, the amount of hydrogen that would need to be carried to the asteroid should be less than a ton if we make efficient use of the energy, as it would be for a mortar of this sort. This would be preferable, in the view of many people, to using a nuclear blast to shove the asteroid out of the way. And exploring this option provides a perfect reason for our next extraterrestrial landing!

We have seen that there is a way to avoid the fate of dinosaurs, if indeed they were killed by a giant asteroid. However, we have not yet considered what happens should a comet, which is what some scientists believe got the dinosaurs, come our way, rather than an asteroid. Comets, for ages considered to be an ill omen, and what we can do about *them* will be considered next.

# 11

# How to Deflect a Comet

The comet, Shoemaker-Levy #9, removed the "giggle factor" that it couldn't happen to us here on Earth. If it [the comet] could hit Jupiter, then one day another one could hit us.

—A scientist commenting on the S/L #9
splashdown on Jupiter in July 1994

Long period comets present a different set of problems than do asteroids. The biggest difference is that there is no hope, using any techniques of science known today, to detect them while they are really far away. They do not generally become visible, even to our most powerful telescopes, until about the time that they cross the orbit of Jupiter. At that point, they are only about a year away from impacting Earth if they are on a collision course. Hence, unlike the asteroids, we have no hope of developing a catalog of every comet that may harm us anytime soon. We shall have to maintain a sky watch for centuries into the future, until some unforeseen breakthrough in science relieves us of that burden. Thus, the time available to plan and execute a comet deflection maneuver will always be short. We will *not* have the luxury of years to effect minute changes in a comet's orbit.

A second major difference is that the composition of comets is very different from many asteroids, as was discussed in chapter 4. This will affect the techniques by which we may hope to move them. A nuclear blast, for example, is much more apt to fragment a comet than an asteroid, since comets are so fragile. On the other hand, a comet almost certainly provides us with a ready supply of gases which we can use for propulsion, if we can learn how to do that properly.

Another difference is that we need to worry about comets much larger than the asteroids we will be concerned with in the near future. After we have a good catalog of the asteroids, the only ones that should take us by surprise are ones that are fairly small. Although these can devastate a city, they do not have the potential to annihilate our civilization the way a very large one could. But a comet of dinosaur-extinction proportions could arrive with no more warning than a small asteroid.

The first law of Arthur C. Clarke, the world's leading science fiction writer, proclaims "When a distinguished but elderly scientist states that something is possible he is almost certainly right. When he states that something is impossible, he is very probably wrong."[1] Further, everyone knows that a committee is more timid and less imaginative than its individual members. Thus, when a committee of distinguished and elderly scientists says that the only way to deflect a comet fast approaching a possible crash with Earth is with nuclear explosives, we should immediately be alerted to cast about for a better solution.

Such a committee of mainly elderly and distinguished astronomers met at Los Alamos in January 1992, at the behest of the U.S. Congress, to consider what could be done about such comets that threaten Earth, once they were discovered. Since Los Alamos is the home of the atomic bomb, and since many of the scientists were drawn from the national nuclear research labs, it is not particularly surprising that the cry came back, "nuke them!" Nuclear explosives dominated the deliberations of that panel. What is rather surprising was the very narrow view taken by that panel in considering the problem. They proposed only Earth-based solutions resting upon current capabilities and technologies!

Is it any wonder, then, that Clarke wrote in *Profiles of the Future*, "Some of my best friends are astronomers, and I am sorry to keep throwing stones at them—but they do seem to have an appalling record as prophets."[2] As further evidence, Clarke proceeds to recount several stories in which astronomers figure prominently as poor predictors of things to come.

## A Look in a Crystal Ball

Let us now consider the following possible scenario for, say, 2064 of what could happen with a long-period comet bearing down on Earth. Because we have projected this scenario so far into the future, it is written in the form of science fiction, but the science is accurate for 1996. In particular, we can calculate that there is enough angular momentum in an asteroid to easily accomplish a comet deflection, provided the momentum is efficiently transferred to the comet.

*The doomsday comet was first seen late Tuesday (Terran GMT) by scout ship 4C-703 of the Deep Spaceguard system, just before it crossed the orbit of Jupiter. That was the primary function of 4C-703, and why it had been placed in a deep space orbit years ago, replacing the B model space watch spacecraft. Station 4C-703 got a few hours' worth of sightings. Seeing the motion of its new discovery against the star background, computers aboard the ship were soon able to predict that this object was potential trouble for Earth. Its data was relayed to Spaceguard headquarters on the hidden side of the moon.*

*Fortunately, the moon was in such a position that the Spaceguard asteroid-tracking laser radar located on the far side could be pointed at the new object. Even though the suspect comet lay well beyond the tracking capabilities of the laser radar, the laser was turned on to illuminate this object in the ultraviolet band. Now, a neighboring picket station, 5C-715, which was much father away, scanning its own sector of space for danger, could turn and see the intruder by the light of the laser. With two watchers, the comet could be triangulated by spacecraft separated by millions of kilometers, and its path predicted much more accurately.*

*Within a day of the discovery, it was clear that this object was a threat. It was not certain that it would impact Earth, since comets are, after all, celestial bodies with built-in jet propulsion, and they can veer erratically in any direction at unexpected times.*

*When this conclusion was reached at Spaceguard headquarters, a Stage One alert was called. The public was not yet notified of possible danger, but all of the assets of the Terran Federation, on Earth or off, could be called upon as required to track down the newcomer and determine the orbit with enough precision to see if a higher level of alarm was justified. At the same time, the defense team went to full alert. They began to examine all of the possibilities for mitigating a disaster of possible catastrophic magnitude. In particular, the Spaceguard computers began to search through the database of asteroids to find one which could be used in this crisis.*

*By Friday morning (Terran GMT), it was concluded that a full alert should be ordered, bypassing two lower levels of response. The order went out to miners, prospectors, and others in or near the asteroid main belt to converge on object 8927D, since it appeared to have all of the properties that would be needed in the interception attempt. Anyone who could get there within three weeks was ordered to move to that destination at maximum speed. Never mind the reaction mass, refueling tankers would meet the assembled fleet at 8927D, and no one was going anywhere else until this threat to Earth had been dealt with. The Spaceguard command cruiser* Clyde Tombaugh *(named for the the discoverer of Pluto) was being dispatched, since it was the closest to the scene. It was expected to arrive in just under two weeks.*

*With two hours, "Iron John" called in from the* Skip-to-My-Loo *that he was only eight hours away from 8927D. He had been prospecting in that section of the asteroid belt. John had earned his nickname because it was said that he could "smell" a workable iron deposit quicker than most survey instrument suites could work up their scientific analysis.*

*When John arrived, he placed his prospector spacecraft into a tight orbit around 8927D, so that tracking data could soon determine the mass of the body. Although it had been cataloged well enough to be readily found for rendezvous, it had never been surveyed at close range. There were just so very many objects this size in the belt! It took almost a day before the tracking confirmed that, indeed, this asteroid had the mass needed for the attempt to deflect the doomsday comet, which by now had been dubbed* Nemesis.

*John's report to Spaceguard Command was laconic. "Well," he drawled, "there is some iron in that sucker, but I sure wouldn't want to mine it." This was followed by the readouts from his survey instruments, which were not as terse, and added a great deal to the knowledge of the rock selected to be the savior of Earth. In fact, the spacemen (and women) had already begun to refer to asteroid 8927D as* Savior.

*The data pouring in from Iron John's ship would be used to make a preliminary determination of the method to be used to nudge the asteroid onto a collision course with the comet* Nemesis, *and to confirm the decision that this was indeed the best way to deal with* Nemesis. *Meanwhile, tracking information about* Nemesis *was confirming that it was going to come much too close to Earth for comfort. Its size was slowly being teased out of the observations, but that was still uncertain, since it was still too small to be resolved into a disk, and there was nothing close enough to experience any perturbation from the mass of the comet.*

*By Thursday, nine days after the detection of* Nemesis, *a small but rapidly growing crew of men and women had arrived on* Savior, *and had a detail map of the surface, completed sonic soundings of its interior, and had the results of a number of bore holes. No previous asteroid had ever been more thoroughly studied. The computer had determined that an addition of eighty meters per second of velocity to* Savior *would put it into position to impact* Nemesis. *It had further been concluded that the best way to add this velocity to* Savior *was with a pair of rail guns. These devices had been invented nearly a century ago to destroy incoming missiles aimed at the United States of America. (The use of a similar technique as a means to move asteroids had been pointed out a century ago.)*

*On Friday, a ten-megawatt powerplant arrived, and mining operations began. The location of sufficient aluminum to make two augmented rail guns had been determined during the previous surveys. Given the high vacuum at* Savior, *and the relatively small volume of aluminum that was required, there was "enough vacuum" to process the aluminum to very high purity before the outgassing got too bad. By Sunday, the first section of rail was in place, anchored to the core of* Savior, *and a cheer went up over the intersuit radio channel.*

*In three weeks, both rail guns had been completed, and the power conditioning needed to transform the powerplant output into pulse power for the rail gun was ready. It took two more days to finish the machines for continuous mining and shaping of pellets for the rail gun projectiles. Then the steady firing of the rail guns began. It took almost all of the first day to remove the spin from the asteroid, so that both guns could proceed to fire as rapidly as stone pellets could be supplied. Little by little,* Savior *began to move in the direction needed to avert a terrestrial catastrophe.*

*Three months later,* Savior *had been moved into a new orbit. Over 4 percent of the asteroid's mass had been hurled out of the muzzles of the rail gun. Now the asteroid would impact* Nemesis *in less than a month, provided only that the comet did not make a large and unanticipated maneuver. That was not expected this far from the sun, although the comet had begun to grow a small tail, indicating that it was a "dirty snowball" and not a rocky body, which would have been more massive and harder to deflect.*

*The miners and prospectors began to dismantle the expensive equipment, such as the power plant and the electronic controls for the rail gun. The rails themselves would be left, since they were too large to move easily and quickly away from* Savior. *Besides they were nothing but high purity cast aluminum, a material quite common in the main asteroid belt. The ships that had brought the crews and supplies to* Savior *were refueled from tankers chartered by Spaceguard Command.*

*As the day of impact approached, the space cruiser* Tombaugh *stood off one million kilometers from the comet, far enough away, its captain hoped, to avoid any of the debris resulting from the collision. Any piece big enough to hurt his tough ship would also be big enough to track on radar. If too many chunks came his way, his astrogation computer might be hard pressed to dodge all of them in the short interval available, but he and the crew of* Tombaugh *had not joined Spaceguard without expecting some danger. His TV cameras were trained on the predicted impact site, and— he knew—billions of TV sets on Earth were tuned in to his signal.*

*His biggest fear, shared by everyone Earth, was that not very much would happen. If the comet were too soft a snowball, the asteroid could punch right through it, like a straw through an Italian ice, without disturbing its soundness or direction very much. Then they would have to resort to more drastic and dangerous methods to save the Earth from a monstrous hammer blow. After all, this deflection technique was just a theory; it had never yet been tried. But it should work; it had been calculated that there was many times more angular momentum in the asteroid than needed to deflect the comet. If only a small percentage of the momentum was transferred to the comet, Earth would be saved.*

As mentioned, this story, although fanciful and laced with science fiction jargon, is based upon real engineering analysis. It could happen this way—*if* humankind chooses to embark upon a spacefaring future as John F. Kennedy called for in his famous 1961 speech. As we have described before, humankind will need perpetual watchmen, and their machines, to avoid being destroyed by a long-period comet.

## The Home of Comets

Unlike the asteroids, whose positions and orbits can be discovered and cataloged, long-period comets come in from from the Oort cloud, and they only come our way once. We cannot see them until, comparatively speaking, they are "right on top of us" in terms of time to respond. The alternative, to go out to the Oort cloud, count them, and predict their future course, seems much more futuristic than the scenario just described of intercepting them on their inbound trip.

The incoming comets have a trajectory, at the distance of the asteroid belt, which is almost directly toward the sun. The asteroids, in contrast, travel in almost circular paths around the sun; almost perpendicular to the direction the comets travel. The cosmos could not have arranged a more effective means of deflecting a comet than this. All that is needed from us is a little fine tuning to make sure that an asteroid of sufficient size is in a position to hit the comet as it speeds toward Earth. Since only a very slight deflection is needed when the comet is as far from the Earth as the main belt, an asteroid with a mass much smaller than the comet can cause enough deflection to make the comet miss the Earth. In turn, this gives us the possibility of moving an asteroid easily enough to do the job.

Of course, if human beings do not choose a spacefaring future that would assure the presence of hundreds of men and women in the asteroid belt by the time of the coming of the comet, we will have to resort to nuclear explosives for deflection. This possibility has been carefully studied by the weapons scientists who met in Los Alamos in 1992 and in subsequent meetings. The conclusion which seems to be accepted by most weapons experts is that the best way to move a comet with explosives is to use a neutron bomb, rather than an ordinary hydrogen bomb. A neutron bomb would be much more "gentle" than a hydrogen bomb, which may sound silly, since we are talking about powerful nuclear explosives. Let us explain what we mean.

As was explained in chapter 10, the energy from a nuclear bomb in space is released in the form of radiation. The capture of some of this energy by an asteroid (or comet) is used to produce a jet propulsion effect to deflect these cosmic bodies. A hydrogen bomb releases much of its energy as "soft X-rays" (that is, X-rays of very limited penetrating power) so that only the outer surface of the threat object is heated. This results in a rather sharp blow to the comet, which may cause it to break into many parts, most of which are *not* deflected away from Earth. These parts may well be be more damaging to Earth than one larger blow.

A "neutron bomb" (technically known as an "enhanced radiation weapon") puts more of its energy into neutrons. These are much more penetrating, so they go deeper into the comet and heat more matter, but to a lesser degree. The result is a more gentle shove to the comet which is less apt to fracture it. This was one of the conclusions of the 1992 Los Alamos deflection conference.[3]

In fact, this conference gives some handy rules of thumb for estimating the amount of deflection velocity that would be provided for various sizes of comets and bombs. For a one-kilometer comet (one big enough to kill a large fraction of all persons living on Earth) a twenty-megaton bomb would deflect it by about two meters per second or about four and a half miles per hour. If this impulse were applied in just the right direction two months before the impact was to happen, the comet would be deflected about one and a half times the radius of the Earth.[4] For a comet, this would insure a miss, even allowing for the focusing effect of the Earth's gravity, which will tend to pull a "near miss" into a direct hit. (For a slowly approaching asteroid this may not be enough deflection, because of its lower speed, thus allowing the object of be attracted by Earth's gravity. A factor of two—or a bit more—must be allowed for these COTEs.)

As we demonstrated in chapter 6, at two months, a comet will be about 1.5 times as far from the sun as from the Earth (1.5 AU) and will be about 56 million miles from Earth, so that Earth will be in no danger whatever from this bomb explosion. We have also showed that we can get a spacecraft out to 1.5 AU in about three months by using a rocket made from parts of a standard communications satellite launcher. So it does appear that we already have in prospect a viable comet defense, even in the short term, if we can just persuade our governments that we want to be protected from cosmic-caused global catastrophe. But we do need an early warning system in place, much like that outlined in chapter 6, together with the means of deflecting comets and asteroids once they are detected.

We must note that the ideas presented here are based upon very incomplete knowledge about comets and their properties. All of our observations are based on measurements taken from great distances (except for the spacecraft that flew through Halley's comet). We urgently need to actually land a probe upon a newly discovered comet and take its measure, so that when one comes our way we will have a body of knowledge prepared to counter it. This will not be easy, since comets move very swiftly, and we will have to treat them as targets of opportunity, since we will not have time to discover one, and then design a probe specifically to examine that particular comet. However, we have enough knowledge to design a comet lander now and have it in readiness when a suitable comet appears.

We spend billions of dollars on projects far less vital to our future than this, but it will take a lot of effort to persuade people of this danger to the point that governments will be willing to act in harmony to prevent these potential cosmic disasters.

## Notes

1. Arthur C. Clarke, *Profiles of the Future* (New York: Holt, Rinehart and Winston, 1984), p. 29.

2. Ibid., p. 22.

3. "Deflection and Fragmentation of Near-Earth Asteroids," in the *Proceedings of the Near-Earth-Object Interception Workshop* (G. H. Canavan, J. C. Solen, and J. D. G. Rather, eds. [Washington, D.C.: GPO, 1993], pp. 89–110), suggests that the fractional areas of the asteroid would be 29.6 percent of the total area, and that the affected volume would be twenty centimeters thick. This material will be blown off the asteroid at high velocity to provide a substantial deflection velocity to the asteroid. Since the "kick" is applied over a much larger surface area, and there is little hammer blow, this is expected to move the asteroid with much less risk of fragmenting it.

4. To achieve a deflection of 1.5 times Earth radius (9,567 km) at 2 meters per second will take 55.4 days, using our rule of thumb, which is probably adequate in this case.

# PART THREE

## COLONIZING THE
## MINOR PLANETS

# 12

# The Road to Space Needs to Be Paved

Where there is an open mind, there will always be a frontier.
—Charles F. Kettering

You may have been skeptical about some of the schemes proposed in chapters 10 and 11 for deflecting comets and asteroids that are about to collide with Earth and cause a catastrophe. "Those things are just not possible," you may have thought. Given the present state of space launch technology, you are absolutely *right!* Well, if those chapters were a problem for belief, you will have real difficulty with some of the ideas still to come in this book, unless you understand that vastly better (and cheaper) space launch systems are not only possible, but long overdue.

## The Evolution of the Space Program

In this chapter, we will describe what is wrong with our present launch systems, describe some that are potentially better, and outline some of the things that are being done to make space travel affordable. Until the present launch systems, all of them, worldwide, are replaced by better ones, we humans will be limited to sticking our toes into the new ocean that

169

John F. Kennedy described in his plea to go to the moon. The road into space needs to be paved!

The very first point that needs to be made is that it is not written in stone that it must forever be difficult and expensive to go into space. Apologists for the current feeble excuse for a space program point to the "vast energy" requirement involved in getting from the surface of the Earth into even low Earth orbit. Anyone who makes this argument simply has not done his or her arithmetic well. The energy contained in a pound of mass in low Earth orbit is less than five kilowatt hours. That is the amount of electricity needed to burn a 100-watt bulb for about two days. That is a lot in comparison to the kinetic energy of a pound of mass moving at 550 miles per hour in a jet airplane,* but on an absolute scale it is quite modest. The retail cost to buy this much electricity may be as much as half a dollar, but the cost of the fuel required to generate it is less than a nickel. The energy requirement is *not* why it currently costs about $15,000 per pound to launch a pound into orbit using NASA's space shuttle!

The reasons for the present high cost of space flight have their origins in the Cold War, which spawned the "missile gap" and then the "space race" between the two major superpowers.

First came the "missile gap." Remember the 1960 presidential contest between Kennedy and Nixon? The presumed fact that the Soviet Union was ahead of the United States in producing intercontinental ballistic missiles was a major issue of debate during that election. The perceived need for the United States to "catch up" with the Soviets meant that our long-range rocket program proceeded at a frantic pace, with cost as no object. Thus, the U.S. rocket program was born as a government-funded program without any serious concern for cost. We believed that our national survival depended upon having long-range, nuclear bomb-carrying rockets at least as soon as the USSR had them in any significant numbers. It turned out that we had learned to build smaller bombs than the Soviets, so our missiles were less capable when it came time to harness them as launchers to embark upon the new ocean of space.

The second major contest between the superpowers was the "space race," which was initiated in 1957 when the Soviet Union launched the first artificial satellite. President Eisenhower had announced that the

---

*The energy in one pound traveling at 550 mph equals 0.0038 kilowatt hours.

United States was going to launch a small, basketball-sized satellite to make scientific measurements of near-Earth space as part of the International Geophysical Year.* The Soviet satellite was several times the size of the planned U.S. *Vanguard* satellite. Moreover, the Soviets succeeded in launching a second, even larger one, carrying the dog Laika even before we had our first basketball in space!

Well, now! The American public felt challenged. How could this backward, communist nation challenge the United States in our strong suit, science? We felt compelled to prove to the world that we were the best in science, industry, education, and economic system. Otherwise, the world might believe that the communist system was better and turn to our opponent for leadership. Thus, the space race was born, and again the perceived need was to win, to surpass, no matter what the cost.

To engage in this race on our behalf, the National Aeronautics and Space Administration (NASA) was created to run our civilian space program and show the world that the United States was not to be beaten by a communist state in any field of endeavor. After Yuri Gagarin became the first cosmonaut to orbit the Earth in 1961, John Kennedy cast about for a goal that would prove this point decisively. His advisors told him that the best goal we could hope to set, and still be sure of beating the Russians, was a landing on the moon, and safe return. Thus, President Kennedy made his speech calling upon America to race to the moon. In the early 1960s, NASA was a young agency, and it attracted the best and the brightest to its challenge to beat the Russians. And beat them we did with a successful *Apollo* lunar landing in July 1969.

After the *Apollo* landing, the public lost some of its interest in space. Richard M. Nixon, in particular, who was president at the time, had little interest in what he perceived to have been primarily a Kennedy program. He had other fish to fry. Mainly, he wanted to extricate the United States from a ruinous war in Southeast Asia. That war, Vietnam, was dividing the nation, and using vast resources, including armies of able, young Americans killed or captured. So NASA and the space program were largely left to fend for themselves.

At this point NASA began its evolution into "just another" vast fed-

---

*The International Geophysical Year was an eighteen-month period between July 1, 1957 and December 31, 1958.

eral bureaucracy. Many of the brilliant engineers who had been vital to the *Apollo* program left the agency, and others settled in for round after round of budgetary wars, trying to get funds to conduct the programs that they felt were worthwhile. Foremost among these programs was the space shuttle, a spacecraft which would be usable for multiple missions. The argument was often heard; "If we had to throw away the jet airliner after every trip to Europe, not many people could afford to go there." Such logic was compelling. If only we had a reusable space launcher, space would be opened to travel for everyone. And so Nixon approved the development of the "reusable" space shuttle.

There is a very long and sad story of what happened to the space shuttle on the way to being the DC-3 of space. It will suffice to say that the space shuttle did *not* become the great doorway to space that everyone wanted, and that many of us expected. Mainly, this is because the space shuttle is *not* "reusable," it is simply "salvageable" after every flight. All of the major components (i.e., the shuttle orbiter and the two strap-on tanks) except the external (main) tank are recovered, but there is a great deal of work involved in disassembly, cleaning, checking, reassembly and testing before that hardware is flown again.

After an airliner lands, any serious malfunctions which were reported by the flight crew are corrected. This can usually be done while fuel is being pumped into the empty tanks, and the cabin is cleaned and resupplied with food and drinks for the next flight. This is typically done by a ground crew of half a dozen people. Then the crew and passengers board and the flight leaves, usually less than two hours after landing. For the shuttle, the time between flights is months, and a ground crew of tens of thousands is involved in preparing the machine for another flight. This is *not* routine airlinelike operation. The total cost for each shuttle flight is impossible to pin down exactly, not only because of the strange way that NASA keeps its financial records, but because ideas differ on what items to include in the cost of each flight. However, the cost for each flight is at least half a *billion* dollars. (That is not a typo, that is 500 million dollars.) According to some consultants, the number is closer to $750 million. And the cargo, including astronauts, carried into space is only fifteen to twenty tons. At these prices, we will never establish a spacefaring civilization.

What happened to the dreams of the 1960s for space travel on a large scale? And why doesn't somebody do something about this deplorable situation?

To answer that, let us look first at other countries around the world. Many countries have begun space programs of their own. The world has decided that space is too important to allow any one nation to have a monopoly on space travel. However, the myth has been perpetuated that space flight is so expensive that only big governments can afford it. This myth has been put forward by those who believe that big government is the way that things should be done. Every country that has a space program runs it through a combination of governments providing the money and private industries (or design bureaus in Russia) doing the work. By maintaining the myth that "space is hard," private investors have been scared away from trying to implement space programs for profit—except for communications satellites. Reversing this trend would mean that, for the first time ever, the cost of getting into space would become a vital factor in the equation.

These ideas have not been disputed, in the United States or in other market economies, because the companies in the private sector who are involved in space flight are making very nice profits from their government contracts to build and operate the present space launchers. It is not nice to quarrel with your present customers, especially when you do not have the vision to see any other way to make money from the business you are in. So the space launch companies are quite happy with the status quo. As long as they can keep selling their present products and services, taking almost no risk, and making nice profits, nothing will change.

So why doesn't Congress, the people's representative, step in to change this? Because the congressmen and -women who fund the national space program are also happy. First of all, almost none of them is technical-minded enough, or imaginative enough, to see how very different things might be. Many of their constituents are very happy with the way things are; space provides a lot of jobs in their congressional districts, and profits to a number of their political contributors. Why mess with a good thing? Space has become a new form of "political pork," just as highways and dams have been for decades past.

Changing the present way of doing space business is going to require some major changes in viewpoint. A few years ago, the Strategic Defense Initiative Office (SDIO, the Star Wars team) had arrived at an interim solution for protecting the United States from nuclear-tipped ballistic missiles that might be launched by our adversaries. Implementing this solution was going to require many tons of satellites to be launched into orbit. Since they had been charged by Congress with putting a missile defense system in place, the Star Wars engineers began to look for ways that they could launch the many tons of satellites required, without requiring a major increase in the National Defense budget. So they asked the American space industry to bid on building a new launch system that would cost far less to put satellites into orbit. One appealing concept that had been advanced was to build a single-stage-to-orbit (SSTO) launcher; this would avoid all of the throw-away aspects of the present space shuttle. Since everything that went up came back in one piece, just like an airplane, it was reasoned that such a space launcher could be operated just like an airplane, and at far less cost. Bids for a new system were requested on such a basis. But a part of the contract was also to be to make the system easy and cheap to operate, dramatically reducing the total system launch cost. All of the major aerospace rocket launch contractors in the United States responded to this offer of a contract. They *all* said it could be done!

Only one contract was awarded, and that went to McDonnell-Douglas Aerospace. They named their concept Delta Clipper, after the name of their existing rocket launcher and the fast Yankee clipper ships of ocean going fame. The first orbital test flight vehicle was designated the DC-1, and the planned progression was that the first operational orbital version would be the DC-3. This machine has now completed its first series of flight tests. Whether they were successful or not lies in the eyes of the observer. Many people, especially those who feel that their jobs (and/or profits) are threatened by a successful new space launcher, will claim that the tests don't prove a thing. After all, the machine did not reach orbit. (This was never intended, and was not within the technical parameters of the design.)

The supporters of the DC-X point to the fact that it did have two major accomplishments.* The first was that it went from initial contract to first flight in twenty-four months, and for a budget of only $65 million. This kind of performance has been unheard of in the aerospace industry

---

*The DC-1 never got beyond the drawing-board stage. The DC-X flew first, then the DC-XA.

since the early pioneer days. The DC-X proved that a small, dedicated team of engineers could create a design in short order, using the ordinary tools of engineering. To accomplish this, they also used a lot of existing hardware from airplane and rocket programs. For example, the flight control computer from an airplane was reprogrammed to control the DC-X. The second accomplishment of the DC-X was complete launches of the same spacecraft only a few days apart, using a ground crew of twenty. There was no need to tear the machine apart, as is done between shuttle flights, to inspect it in excruciating detail! A rocket *can* be flown just like an airplane!

The initial reaction of NASA was that this demonstration was not important, and that building a useful SSTO launcher was years, if not decades, away from practicality. However, a lot of space enthusiasts believed that something valuable had been accomplished. The old paradigm was challenged by this upstart rocket. The activists wrote to Congress, and generally made a lot of noise. Since they had the facts on their side they were hard to ignore. It was true that every major U.S. aerospace company has told the SDIO that they believed that they could build a successful SSTO, and that this would save a lot of launch cost and open a new frontier to human expansion. The result was that NASA was compelled to take the new concept seriously; Congress put funds into the NASA budget for new launch technology and insisted that NASA explore the possibilities of reusable space launcher vehicles again.

Several things are happening in NASA as a result of this pioneering effort. First of all, NASA has taken over the original DC test vehicle from McDonnell-Douglas, which finished with the test series it had contracted with Star Wars to perform. NASA is going to use the vehicle to create new technology that will advance the date when a practical SSTO can be flying. In particular, DC-XA (as it is now called) was outfitted with new propellant tanks made of new, lightweight materials. This is a start toward meeting the objection that the DC-X could be flown like an airplane because it was built like a battleship. Actually, it was built of available, low-cost materials and technologies to save time and money and because the issue of how much it weighed was not a concern in the initial flight demonstrations. However, it is true that to be a successful SSTO launcher, the craft must be of much lighter construction. The DC-XA had four successful flights in the spring and summer of 1996. Each flight went higher, farther, and faster than the one before, adding even more proof that a rocketship *can* be flown like an airplane. As a result, there is now little question that a space shuttle replacement can be built that will fly much more cheaply. Sadly, on the last planned flight of the DC-XA, one landing gear did not extend on landing, and it toppled over and burned.

As a result of the success of the DC-X (and pressure from space activists), NASA initiated two new rocket development programs. These are known as the X-33 and the X-34. The X-33 is intended as a technology demonstration program for a "large" reusable space launcher, probably a single-stage-to-orbit. Three contracts were awarded by NASA in January 1995 for the initial design studies of the X-33. These were "cooperative development" agreements between NASA and the aerospace contractors. Both parties invested some money, and the aerospace companies conducted the work, using their own designs. These studies were completed early in 1996, and the three industry teams made proposals to NASA to continue the development into a flight test stage. Rockwell proposed a concept that looked very much like a fat shuttle. McDonnell-Douglas proposed a derivative of the DC-X that took off and landed vertically. Lockheed-Martin proposed a design based upon the HL-10 lifting body, but with a new design of rocket engine that looks like a fat delta-winged airplane. It takes off vertically and lands horizontally, just as the space shuttle does.

These proposals were evaluated by NASA, and a contract was awarded to Lockheed-Martin for their design. The intent is for the government to fund the flight test program that will prove the feasibility of the concept. Then the plan is to have private industry to take over the operational phase of the program; that is, to design, build, and operate spaceliners just as industry now operates airliners. There may be a contract guarantee from the government to provide a certain level of launch business, just as the airline industry got started with air mail subsidies.

There are many design concepts that can be used to provide a fully reusable space launcher. The first design choice is whether to use a single-stage or two-stage launcher. (The former has a single rocket booster while the latter employs two boosters, one on top of the other, in a multistage fashion.) Almost all space launches to date have used at least two stages. (The notable exception was the Score satellite, which was basically an Atlas ICBM with a tape recorder and a transmitter. It transmitted greetings to the world from President Eisenhower during the Christmas season in 1958. This was largely regarded as a publicity stunt, since the useful cargo was so small compared to the size of the launch rocket.) The preferred idea is a single stage, all of which returns ready for reuse, just like an airliner. There are many reasons for this preference. One is that since no parts are dropped off during launch, the spacecraft can be launched from anywhere. Otherwise, the rocket path must be across an ocean or large uninhabited wasteland. But with an SSTO, your city can have a spaceport! The other reason is that, since there are not two stages to be mated and "checked out," it is faster and much cheaper to just refuel and fly again. There are many reasons to expect an SSTO to be cheaper to operate than a two-stage-to-orbit, often called a TSTO.

For these reasons, the development work on the X-33 is being concentrated on the SSTO concept. Within this approach, there are still several major design choices. Simply they are:

- Vertical launch, horizontal landing. This is what the shuttle does.

- Vertical launch, vertical landing. This is what the DC-X did.

- Horizontal take-off, horizontal landing. This is what airliners do.

- Horizontal take-off, vertical landing. Almost nothing ever does this, and no-one is presently advocating this for near future space launchers. It is hard to see much advantage to this mode.

Besides the X-33 effort, NASA is working on a program called the X-34. This is a prototype demonstration for a small, partly reusable, launch vehicle. One small aerospace company, Orbital Sciences Corporation, in the 1980s developed a small launch vehicle with a twist: it was launched by dropping it from an airplane in flight. This approach, called Pegasus after the flying horse, has several significant virtues: First, releasing the rocket at high altitude and at jet plane speed gives it a valuable head start. In particular, the rocket engine (which works best in a vacuum) works better at high altitude than at sea level. Second, releasing the rocket from an airplane *requires* that the launch crew be small; there is not the expensive cast of thousands usually associated with the launch of a big rocket from the Kennedy Space Center, for example. Third, the launch site is portable. This eases many of the restrictions of launching from an established site, and offers important advantages to the customer for the launch services.

This concept has led to a the successful establishment of a new launch company in the face of major established competition. The X-34, which may be considered an "evolved Pegasus" is intended to replace the throw-away first stage of the Pegasus rocket with a fully reusable stage. The X-34 will carry an upper stage to the fringes of space to continue on its way, while the X-34 returns to Earth to be refueled and to fly again another day. (The upper stage is not intended to return to Earth for reuse.) This should make space even more accessible for small satellites, such as those that universities would like to launch as a part of their science research programs.

There are other space launchers being developed, each with its own reason for being. Lockheed is developing a small expendable solid rocket called the Lockheed Launch Vehicle (LLV) to launch satellites for private ventures. One major target market for both Pegasus/X-34 and the LLV are small communications satellites.

There is also a completely private venture, Kistler Aerospace, developing a fully reusable launch vehicle. They are planning a single-stage-to-orbit with a "launch assist platform," or LAP. The LAP is a specialized rocket that some people call a "cheater stage." It will take the rocketship (as Kistler calls it) to high altitude and a high subsonic velocity for launch. This concept has several advantages over the others being developed.

First, just as for Pegasus, it is easier to get to orbit from a high altitude and high speed start, so it should be much less costly to develop than a completely single stage design. This is especially important for a small, start-up venture such as Kistler Aerospace, which has limited funding.

Second, the LAP is very easy to develop. It does not have to go very high or very fast, so it does not need parts that can withstand the space environment, or the searing heat of re-entry, as most first stages do. Further, it should be cheaper to fly than a converted airliner, such as that used to launch Pegasus.

Third, the LAP will land, just a few minutes after launch, right back at the starting point of the flight. It acts more like an elevator than an airplane or a rocket. This will make it easy to use one LAP often to launch many orbital rocketships, since it will always be at the launch site.

Many experts believe that, as a result of these new programs, the next generation of space launchers can be ready for flight in less than ten years and that the cost to fly to space can be reduced by a factor of ten by that new generation of launchers. Compared to today's $15,000 or more per pound in orbit, there is every reason to expect that this cost can be reduced to $1,000 or so in less than a decade. And that will not be the end of cost reductions. Just as in the airline industry, we have every reason to expect that space flight will get progressively better *and* cheaper as we begin to develop it. Some experts predict that the cost of a trip to orbit should cost no more, for our children, than a round trip to Australia (from the United States) now does. In fifty years, our present symbol of "high technology," the space shuttle, will appear to be hopelessly primitive, just as the Wright Flyer of 1903 now appears to us to be hopelessly primitive.

Of course, we have been talking only about the first step of an interplanetary journey, the trip into low Earth orbit. However, that is the hardest part of the trip. To make that leg of our journey, we need to fight Earth's gravity field and the resistance of the atmosphere. Most of the places in the solar system that we wish to visit do not present these twin obstacles. Of the inner planets, only Venus has much atmosphere, and it is much too hot to visit, at least until we have acquired a whole new set of technological capabilities. Only the giant planets have more gravity than the Earth, and they all have many satellites (moons) that are low gravity worlds. So, we shall not lack tempting targets for our explorers for many years to come.

## Traveling Throughout the Solar System

For moving about in this solar system of low gravity and mostly airless worlds, there are already a number of means that we know we can use, *and we have not even really begun to work on the problem!* We will mention just a few of the concepts that have already been worked out to deal with the problems of transportation in the solar system.

Ion engines are the next new propulsion system that we will use in getting around the solar system. A conventional rocket engine depends upon the energy of combustion to heat gases to a very high temperature after they have been pumped up to a high pressure. These gases then expand through a rocket nozzle to produce thrust, just as the air in a balloon expands and causes the balloon to flit about the room when you release the neck. This process is so well understood that we know that we have just about reached the ultimate performance that we can expect from chemical rockets. To do better, we will need to invoke new principles.

The first step in obtaining higher performance from our rocket-like devices is to separate the energy source from the "stuff" that we squirt out the back end. One way which has been demonstrated is to use an electrical source, such as a solar panel or a nuclear reactor, to expel a separate "working fluid" from our engine to provide thrust. This working fluid may be any of a variety of things, but the current favorites are argon and xenon gases. These gases are ionized by an electrical field, and then the charged atoms of gas are accelerated by magnetic fields to very high velocities. This method has the advantage of producing much more spacecraft velocity for a given mass of propellant. However, there are some major disadvantages as well.

First, this method uses a great deal more energy to produce a given velocity than chemical propellants. As long as this energy is abundant, as from solar or nuclear sources, that is not too much of a concern. However, the real limiting problem is that these energy sources are quite heavy compared to the chemical rocket. Thus, the acceleration of these engines is very low, which, in turn, has two consequences. First, present ion engines are utterly useless for getting off the surface of the Earth, or even the moon. They have to be in space, or perhaps on an asteroid, before they could be used to start a trip. The second problem, of less concern unless you are an impatient person, is that because of this low acceleration, it takes a long time to reach optimal speed. We can deal with this, generally, just by being more patient.

A related concept is the thermonuclear rocket. In this case, the energy source is the energy produced by the nuclear processes in the reactor, and the working fluid is some other substance. In general, we will only be concerned with two working fluids: the best and the most plentiful. A thermonuclear rocket works rather like the chemical rocket, except that the gas is heated by the nuclear reactor, so we can choose whatever gas best suits our purpose. If we want the highest performance, we will choose hydrogen, the lightest gas in the universe. This will give us the highest speed for a given quantity of gas. If we want to use the the material which is already on the rock we want to move, we will get the best results from a given amount of hardware carried to our island in space by using the water we find there (recall that water is quite abundant in comets and asteroids) in a steam rocket.

There is yet another means for getting around the inner solar system, without using any propellant. This method is called "solar sailing." Light exerts a pressure when it strikes something that either absorbs or reflects it. Reflecting is better, since a perfect reflection will exert twice as much pressure as perfect absorption. This light pressure is extremely small. However, it is so noticeable that a light, inflated structure, such as the Echo balloon communications experiment of the 1960s, will be affected by it. Since this pressure is so small, the structure necessary to resist in can be constructed of the flimsiest of materials; very thin films of aluminized plastic can be used to make solar sails. Again, this method is rather slow, but it can be used to move quantities of raw materials about the inner solar system at very reasonable cost.

It is also possible to use tethers to move things without using rockets. Consider a geosynchronous satellite. Because of the laws of orbital mechanics, this type of satellite appears to hover motionless over a single point on the Earth's equator. It would be possible to lower a long and very strong cable from a platform in geosynchronous orbit to the ground. If you also put up a counterweight above that platform, you could use an ordinary elevator to get up to that platform, and you would be in orbit without using any rocket at all! There are two things wrong with that idea in terms of use in 1996 or any time soon. First, we do not have any material strong enough to make a cable reaching all the way from Earth to geosynchronous orbit. However, with the invention of Kevlar®, a material that is both light and strong, we are much closer than we were when this idea was first advanced many years ago. It is impossible to rule this concept out simply on the basis of today's material properties. The second obstacle is the presence of many Earth satellites at much lower altitudes. Do we have to remove them all, or as someone noted, perhaps we can cause the cable to reverberate just as a low satellite comes to it? Whether this idea is ever useful at Earth or not, it certainly may be useful for small satellites, or even small planets. It has been calculated that present materials are strong enough to allow such a tether to be built to get from the surface of Mars into orbit around it. This may not be very useful for the first Mars landing party, but it can certainly affect the commerce between planets in the distant future.

One way that we have already used to navigate in space is by momentum exchange with the planets. To send the probe *Ulysses* close to the sun, we did not launch it directly toward the sun, we sent it *away* from the sun. Our best rockets were not powerful enough to send it into the polar (as opposed to equatorial) orbit of the sun that was desired. So, we sent it to Jupiter. It did a sling-shot maneuver around Jupiter and went into a polar orbit around the sun. (This specific maneuver was suggested by one of your authors in 1963, but it was not done until 1972, when *Pioneer 10* swung by Jupiter on a mission to explore deep space. This proved the feasibility of the concept, and it has been used in many interplanetary missions since that time.) This form of maneuver, which is now called a "momentum exchange" by engineers, has been used by a variety of interplanetary probes, including the *Galileo* mission to Jupiter, and the *Pioneer* probes now leaving the solar system for a destiny unknown.

This method is quite useful, if you happen to have a planet just where you need it. Since that is not always the case, it is worthwhile to know that we can devise momentum transfer devices where we do need them. Such devices are called "rotating tethers," and they can replace vast quantities of propellant in the future colonization of the solar system. Robert Forward describes such devices in more detail in his book *Indistinguishable from Magic.**

The important thing to accept is that we are mere beginners in our voyages to the planets. What we have done so far has been slow, painful, and expensive. However, it is not carved in stone, anywhere, that it will forever be difficult and expensive to move about in our own solar system. As we learn to do this it will become easy and fun. It will become the great adventure of the twenty-first millennium!

---

*(Riverdale, N.Y.: Baen Publishing, 1995).

# 13

# The First Human Visit to the Asteroids

A man's reach should exceed his grasp, or what's a heaven for?
—Robert Browning

One of the current debates about space is the destination and the timing of the next manned flight beyond low Earth orbit. Everyone knows that the moon was the first extraterrestrial body visited by humans, but where to now? Many experts advocate that Mars should be the first planet to be visited and therefore our next destination. Others want to return to the moon to establish a permanent outpost; the beginnings of a lunar colony. But there are also excellent reasons for picking an "island in space" as the first target of such a flight. For one thing, the energy requirements can be quite modest for a flight to an Earth-crossing asteroid. And given their potential for damage to Earth, such exploration seems very much in order.

In earlier chapters we talked about deflecting asteroids and comets before they could do major damage to planet Earth. There we discussed some schemes which assumed that there was already a major human presence in the asteroid belt. That may have seemed like science fiction to you, and at present it is just that. However, in this section, we will talk about how to make such a human presence into reality in the near future.

It has been twenty-five years since humankind last ventured past low

Earth orbit. This seems incredible to those of us who watched the first man walking on the moon. Who could have imagined that we would abandon our dreams for exploring the solar system, and remain content to be confined to our home planet for this long? Since we seem to have abandoned those dreams, the very first requirement for flights to the asteroids (or the moon, or Mars) is a determination to be on the move again, and to take up the cause of human exploration and eventual settlement of the solar system. We will start this chapter by assuming that, somehow, we humans again feel our natural curiosity for exploration and adventure, and decide to venture forth to the asteroids.

The value to science of such a trip should be obvious. Most important, however, would be the value of the trip in scouting the path to the future utilization of the planetoid as a site for a "filling station," as a source of raw materials, as a site for a colony, and as a candidate for possible capture and return to Earth orbit. Quite possibly the second trip to a planetoid would be a capture expedition itself; thus, the high importance of the first reconnaissance flight should be evident. (The technique for capturing planetoids will be discussed later in the book.)

The men and women of the first planetoid expedition would be particularly interested in determining the chemical composition of the planetoid, the water content, the location of high-grade ores, the structural properties (could it withstand the repeated shocks of nuclear bomb blasts used for capture?), the best sites for bases, mining camps, propulsion chambers for capture, and so on. The crew of the first trip might also take the initial steps toward preparing the planetoid for occupation or capture and might hollow out caves for living quarters for themselves and future residents. Thus, we see that the importance of this first flight, as compared to some other interplanetary voyage, cannot be separated from the enormous future potential value of the planetoids. Therefore, the future potential of the planetoids in the different phases of the exploitation of space must be evaluated in assessing the importance of making the first exploratory flight.

The first thing we will discover when we decide to send human explorers to the asteroids is that we do not currently have any space vehicles capable of undertaking such a flight, nor do we have vehicles capable of taking humans back to the moon! This is how great our lapse of interest in space has been. It will be useful to consider a trip to the asteroids in terms of what we know about the trips we have taken to the moon. The recently popular movie *Apollo 13* is a reminder of the machines and techniques needed in space exploration. We will draw upon *Apollo* for examples of what needs to be done, and also for things which will be different for asteroid flights. There will be two reasons for differences. Some things will be different because of the nature of the target and the missions, and others will be different because we have advanced so much in technology since the days of *Apollo*.

## How Shall We Get There?

The first question is what shall we use for a launch vehicle? The big, old, reliable, first generation Saturn V boosters that sent astronauts on their way to the moon are not available anymore. Moreover, the tools needed to build more are gone. Even if we had the tools and the blueprints, we would not want to build any more of these vehicles; they are obsolete. For example, the electronics are much heavier and less reliable than what we would use today. Similarly, the second generation space shuttle booster will not get us out beyond a near Earth orbit. So we are faced with a choice; build a new large launch vehicle, or assemble our deep space spaceships in low Earth orbit.

Before we try to choose between the two, perhaps we should stop and think about what we need to launch. Is an *Apollo*-type spaceship what we need, or do we need something entirely different, such as the new vehicles discussed in the last chapter? How long is the trip? How much propulsion do we need to to get home?

Before we can answer that question, we need to recognize that there are asteroids in two places in the solar system. The vast majority of asteroids are in the "main belt," which lies between Mars and Jupiter, far out in the solar system. However, there is a pesky minority of them which threaten the Earth, as we have described earlier. It turns out that this group of COTEs will be far easier to visit due to both time and propulsion considerations. A trip to these is comparable in many ways to the moon trips with which we are familiar. The rocket to send us on our way will have to be a little bit more capable than the Saturn V, but nothing that our engineers cannot easily handle. Moreover, the trip time can be held to a few weeks, so we can readily carry all of the supplies that our explorers need. Finally, the trip back requires less fuel than a return from the moon.

On the other hand, a trip to the main belt will require a total trip time of several years. This will require a much larger spaceship, with more supplies, and probably a larger crew. If we want to land on some asteroids while we are out in the main belt, we will need something like an enlarged lunar landing module to change our velocity to allow us to do this, and a lunar return module to bring us back again. Since the energy requirements are fairly small for both of these functions, we may be able to combine them into one spaceship, saving both time and money in preparing the trip. However, this mission to the belts will require a great deal more time and expense than a trip to the Earth-crossing asteroids that threaten us. Therefore, in this chapter, we will discuss only the prospects for a crewed* voyage to a COTE. Later, we will talk about options for exploring and colonizing the main belt.

## New Booster Options

There is one option for building a large, Saturn class booster in the near future that has been considered for a long time. This is called the Shuttle Derived Heavy Lift Launch Vehicle (SDHLLV, or sometimes just HLLV). This rocket would use the main propulsion parts of the current space shuttle, the solid rocket boosters (SRBs), the external tank (ET), and the space shuttle main engines (SSME), in combination with a new cargo pod that would replace the shuttle orbiter. Since all of the main parts exist, and have been tested extensively, and used in manned flight, most experts consider this to be an easy way to build a reliable large launch vehicle.

Such a combination could lift about a hundred tons into low Earth orbit, far more than any current U.S. launcher. NASA has considered this option for years, and refused to build it. There is no current need for such a large launcher in the view of NASA management. Many independent engineers consider this rocket the cheapest and fastest way to provide the payload capability to conduct *any* mission beyond low Earth orbit. How-

---

*We use the term "crewed" rather than manned because we have every expectation that, in the future, all long space trips will have crews composed of both men and women. Shuttle experience proves that this works well, so we use the term "crewed" even though it sounds rather awkward.

ever, going boldly beyond low orbit is *not* anything that NASA is presently willing to consider.

There are certainly some other booster options. One that has its advocates is to use the Russian Energia booster developed by the USSR to answer the challenge of the space shuttle. It is reported to be capable of carrying almost 200 tons into low Earth orbit. However, it has only flown once, years ago, with less than complete success, and many experts believe that the few models which still exist are no longer capable of safe flight. However, *it is* an existing design, and it is quite possible that a joint U.S. and Russian program could return this rocket to flight status as a viable option for launching large payloads into space for adventures such as a mission to an asteroid.

There is always, of course, the option to build a new, modern, large launcher from scratch. Such an approach has considerable appeal. We have learned so much about space flight since the last completely new rockets were built that it is tempting to start all over again and use everything we have learned. To do this will take several years and billions of dollars, even if it were done efficiently. And many people question whether NASA can conduct this kind of work well now, since it appears that a heavy layer of bureaucracy has set in since the halcyon days of *Apollo*.

One suggestion for this new approach comes from Dr. Robert Zubrin, a futurist thinker from the Lockheed-Martin company. Zubrin has proposed, among other things, an early return to Mars, using gases from the Mars atmosphere to supply a major part of the return propellants. This approach, called Mars Direct, requires a large launch vehicle such as would be necessary for asteroid missions. He has developed a conceptual design for this launcher, called Ares, which uses mainly shuttle-based hardware, but with a new upper stage. The launch stage (called the zero stage in the rocket business) consists of two advanced solid rocket boosters (ASRB). These are strapped to the external tank, which has four space shuttle main engines on an outrigger. The launcher has a new upper stage to provide interplanetary capability.

According to Dr. Zubrin, this machine has a launch capability of 47.2 metric tonnes* for a Mars mission (C3 of 15),† and 52.1 metric tonnes for

---

*A metric tonne and an American ton are close to the same mass. In fact, the terms are often used interchangeably, as the uncertainty in what we are describing is larger than the small difference between the two terms.

†"C3" is a measure of the energy required by an interplanetary trajectory. The larger the number following C3, the greater the amount of energy required. (For further details on C3, see the note on page 191.)

a lunar mission. We estimate that this would be reduced to less than thirty metric tonnes for a Toutatis mission, since that trip requires a greater launch velocity than a Mars flight (estimated C3 of 63). However, the payload requirement is expected to be considerably less than for a Mars mission.

The other way in which we can build a spaceship to take us to the asteroids is to do this in orbit. As we tried to show in the last chapter, some new rockets are needed to provide cheap access to space (CATS) or we will remain mostly Earth-bound far into the future. Once we have CATS, it is entirely reasonable to consider working in low orbit as a viable means of building things.* Since our asteroid explorer needs several parts, or modules, anyway, these could be built and launched separately, thus minimizing the amount of work required in orbit to prepare it for a voyage away from Earth. Therefore, it may well be cheaper to assemble an asteroid-exploring ship in orbit, using modules built on the ground, than to build it on the ground and then have to develop a huge new rocket to launch it all at once. For our present purposes, we will assume that our asteroid explorer will be built in this manner.

## Other Options

Now we can turn our attention to the other questions we raised earlier. First, let us consider our destinations. Here we have many options. When Cox and Cole wrote *Islands in Space* in 1964, they were hard pressed to find even a single asteroid that would be suitable for a short and easy visit. They proposed a hypothetical asteroid, which they named *Hypotheticus*, to satisfy their need for an easy target to visit. They said: "A landing could be made on [comet] Wilson-Harrington or Hypotheticus for a total velocity change of only 13 thousand feet per second (4 km/sec). It

---

*Low orbit is definitely preferable to "high" or geosynchronous orbit as a place to build explorer spaceships for several reasons:

1. It is much easier to get to low Earth orbit (LEO) from Earth than to get to geosynchronous Earth orbit (GEO).

2. Working at GEO subjects the workers to much more hazardous space radiation than does working at LEO.

3. There is only a very minimal advantage regarding the difficulty of the mission to leave from GEO as compared to LEO. This advantage is offset many times over by the penalty in getting to GEO with the building blocks.

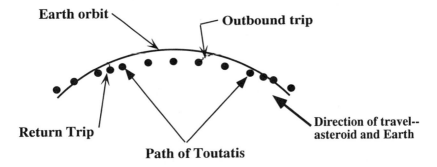

Figure 13.1: Round Trip to Asteroid Toutatis

is reasonable to expect that this object will eventually be rediscovered or that a new object with similar orbit characteristics will be found."[1]

Their prediction that objects meeting the requirements for an easy visit would be found has been proved correct. Wilson-Harrington has been relocated, and new asteroids that will be even easier to visit have been found. One of these is Toutatis, which will make five close approaches to the Earth in the next twenty-five years. We will use this object as our example in discussing early visits to asteroids, for reasons that will become increasingly clear. Appendix D1 contains a list of fifty-nine close approaches to Earth that will occur between 2000 and the year 2020. Hence, there are many other targets of opportunity if we decide for any reason not to visit Toutatis first.

First, let us consider the flight plan for a trip to Toutatis. (Trips to other COTEs will differ in detail, but this discussion will cover the essential points for any similar trips.) Figure 13.1 shows how the orbit of Toutatis crosses the orbit of Earth at two places as Toutatis makes its trip around the sun. We want to launch our intrepid explorers toward Toutatis as it first crosses the Earth orbit. This is shown in figure 13.1 by the lines going from Earth to the orbit of Toutatis. We can then stay on Toutatis for a several weeks as it moves closer to the sun than to Earth. As the asteroid starts on its trip back out to the asteroid belt, our spaceship and its crew will push off from Toutatis on a path that will take them back to Earth. (Since Toutatis is moving faster than Earth, we must play "catch-up" on either the inbound or outbound leg.*) It must be noted that each close encounter with Toutatis will differ in detail, but the general concept will be similar. For asteroids that do not actually cross the Earth's orbit, the path of the spaceship will be different, but the concept of "landing" while the asteroid is close will still apply.

We need to note that both the landing on Toutatis and leaving the asteroid will be much easier than the corresponding events on the moon flights, primarily because Toutatis is so tiny it has almost no gravity. As we approach the asteroid, we will be moving slowly relative to it, since our launch from Earth has put us into essentially the same orbit as Toutatis.

In fact, "docking" is a better term than landing, since the maneuver is much more comparable to the space shuttle engaging with the *Mir* space station than to a moon landing. The principle difference is that Toutatis will not cooperate with us as *Mir* cooperates with the shuttle. In fact, we will probably want to drive pitons (mountain climbers' stakes) into Toutatis to secure our spaceship from drifting off during some minor disturbance. Similarly, as we leave, all we need to do is to give our ship a rather small "shove" toward Earth so that we will intercept our home planet as Toutatis recedes from the sun, rather than accompany it all the way out to the main asteroid belt. But note, we *could* go with Toutatis all the way to the asteroid belt, if only our supplies were adequate.

In guessing how large a spaceship will be required, it will be very useful to draw some comparisons with the size and mass of the *Apollo* vehicle. The command module in which the three astronauts lived during their trip to and from the moon had a mass of less than 13,000 pounds, and the attached service module had a mass of another 13,450 pounds less the propellant needed to enter and leave lunar orbit. Thus, the *Apollo* crew compartments had a mass of about thirteen tons. Now the trip to the moon took three and a fraction days, and the return a comparable time. Add in the few days on the lunar surface, and we see that this mass provided what the astronauts needed for about ten days.

To accommodate the somewhat longer trip to Toutatis, we will have to add more supplies; specifically, more oxygen, more food, more carbon dioxide removal canisters, and the like. This will add another a ton or so of mass. Further, the amounts cited above do not include several hundred

*Toutatis will take 56.9 days from its first Earth crossing until the second, while Earth will take 60.9 days for the same transit.

pounds of oxygen and hydrogen for power and water. The astronauts will probably need a rather larger space to occupy for a trip of a month or more. This is reasonable, since it does not take much mass to enclose extra empty volume for the crew, especially if we leave from Earth orbit. And finally, we may want to increase the size of the crew from three to perhaps five people. This will of course increase the need for food and air.

Offsetting these increases are the advances that we have made in technology since the *Apollo* spaceship design was laid down more than thirty years ago. Two areas in particular will be invoked. We have new materials, specifically, composites such as graphite epoxy, and others such as Kevlar®, that can be used to make structural members lighter. And of course, the electronics can be much lighter and more reliable than the primitive computers and stable platforms of *Apollo*. Referring to *Apollo 13*, there will be no fear of the dreaded "gimbal lock" mentioned in that movie, since modern guidance systems do not use stable platforms such as that *Apollo* had.

We did not undertake the conceptual design of an asteroid-exploring ship for this book, so we cannot say with high confidence what its mass will be, but based upon these kinds of considerations, twenty tons seems like a reasonable estimate for our initial exploration vessel. This is the mass that must be pushed away from Earth at a velocity great enough to reach Toutatis, either by a large rocket launched directly from Earth, or from modules and propellant assembled in Earth orbit.

## Barriers to Be Faced

In discussing getting around the solar system, the twin measures of difficulty are propulsion requirements and total trip time. The propulsion issue is simply one of the mass of fuel required to move the spaceship around, and this is described by the parameter called Delta-V (for change in velocity).* The trip time, of course determines how long the crew will

---

*Delta is the Greek letter universally used by engineers to describe "change in"; "V" stands for "velocity." Since the amount of propellant required is exponentially related to change in velocity, Delta-V is universally used as a measure of space mission difficulty. Among interplanetary mission planners, the square of the Delta-V above Earth-escape velocity has a special name: it is called C3, and is a direct measure of the energy involved in conducting a mission. Any technical literature about planetary flight will make frequent reference to C3.

be away, and how much food, water, and air they will consume. Long trip times, of course, increase the risk of medical emergencies for the crew, as well as the effect upon their emotional well being.

The Delta-V requirement for the departure to Toutatis will be about 13,000 mph (5.75 km/sec) over the velocity in a 500 kilometer orbit. This can easily be achieved by an upper stage rocket of conventional design, since this is just over half the Delta-V required for SSTO. Note that very little velocity will be required for the return trip, provided that an Earth-entry trajectory is used to slow the returning vehicle, in a fashion similar to what was done for *Apollo*. It should be noted that the total Delta-V required from the *Apollo* spaceship to enter lunar orbit, land, and return was 10,500 mph (provided by three rocket stages), or almost as much as that required for our asteroid mission. We can probably provide all of the velocity requirement with a single propulsion stage, rather than the three required by *Apollo*. This would make an asteroid trip much cheaper than an *Apollo*-type return to the moon.

Based upon this Delta-V estimate, and our guess of twenty tons (40,000 pounds) for an asteroid spacecraft, then the mass needed in low Earth orbit (LEO) to perform this mission will be about 175,000 pounds. This could be provided by nine flights of the Delta Clipper SSTO proposed by MacDonnell-Douglas. (That launcher is projected to carry ten tons into LEO, for an eastern launch.) The first two launches would provide the spaceship and its orbital booster, and the remaining seven would provide all of the supplies and propellants. Assuming that the goal of $1,000 per pound in orbit for early SSTO operations is met, then the launch cost for this mission will be about a third of the cost to launch one space shuttle! This is surely a bargain, since there is so much that we need to learn about these threats to our survival.

As a basis for comparison, table 13.1 shows the velocity change (Delta-V) which would be needed for some typical lunar and interplanetary flights. Note that all landing trips assume a return flight as well and that all values are for minimum energy trips. To make the trips in shorter times, higher velocities are required. We should point out that the Delta-V required for a single-stage-to-orbit launch is close to 9.75 kilometers per second (32,000 feet per second). Thus, once we attain this goal, these other missions look much less demanding.

## Table 13.1: Approximate Space Transportation Velocity Requirements

|  | miles per hour | km/sec |
|---|---|---|
| Low Earth orbit to escape | 7,500 | 3.20 |
| Low Earth orbit to high lunar orbit | 8,200 | 3.66 |
| Low Earth orbit to Mars fly-by | 8,900 | 3.96 |
| High lunar orbit to lunar surface | 5,500 | 2.44 |
| High lunar orbit to Mars orbit | 5,500 | 2.44 |
| Surface of moon to Mars orbit | 10,250 | 4.55 |
| Surface of Mars to low Mars orbit | 10,250 | 4.55 |
| Low Mars orbit to Earth | 6,900 | 3.05 |
| Low Mars orbit to planetoids | 10,250 | 4.55 |
| Planetoids to Jupiter | 6,900 | 3.05 |

A minimum-energy round trip to the surface of Mars might be made for a total of 24,000 feet per second if the atmosphere of the planet can be used to decelerate the spacecraft before landing and if the Earth's atmosphere can also be used for braking on the return flight.

We want to note that there is a very good prospect that, unlike *Apollo*, we can save all of the hardware used for each flight, so that we can easily afford to make repeated trips to the COTEs that have already been identified. Remember that, on average, we will have several opportunities each year far into the future. Rather than separate a "command module" and throw the rest of the ship away, we can bring the whole thing back into low Earth orbit by using atmospheric braking. We can dive through the upper atmosphere and reduce our excess speed such that a minor boost from our engines will put us back in our starting orbit. The crew can then return home as passengers on a returning SSTO flight. The spaceship can be checked out in orbit, refuelled, and be ready to go to take a crew to another asteroid in a matter of days or weeks. In this manner we can rapidly learn what we need to know about our close-approaching neighbors.

Given the sad state of our space program at present, what new things do we need to move forward toward a vigorous effort to move into the cosmos, and to defend our planet? First, we will discuss the technology needed in the next few years in order to get moving again.

We have already noted that we must have a much cheaper way into orbit than the present space shuttle provides. It would also be useful to have a heavy lift launch vehicle, although as noted, we *can* conduct such missions using spaceships assembled from cargo brought up by smaller launchers.

We need better means of protecting astronauts from space radiation. In particular, solar storms have the capability to kill a crew in minutes in a ship such as *Apollo* that has essentially no radiation shielding. This is not a problem for low Earth orbit, since the Earth's magnetic field deflects the charged particles so that we can operate successfully for years in low orbit. Once we are away from the Earth's magnetic field, the situation changes dramatically. On *Apollo*, we took our chances and won; the crews were not injured by space radiation. However, as our trips venture farther from Earth and take more time, we will need better solutions to the space radiation problem.

We also need to develop the hardware necessary for a long duration flight. The space habitat we need for a trip to the asteroids is similar in many respects to what we need for the several-day trips to the moon, and for trips to Mars and elsewhere in the solar system. In particular, we need to develop improved life support systems. For very long trips, we will want to recycle our water and air, rather than just carrying ever larger supplies. This principle has been discussed for years, but NASA is not doing anything about it. We will also need to replace the fuel cells that power the shuttles with either solar or nuclear-powered supplies. It would be easiest to replace them with solar power for ships, such as the asteroid explorer we have been describing, that leave from and return to orbit. Of course, these solar panels must either be retracted for the atmospheric braking maneuver, or perhaps just discarded and replaced after each flight.

In short, there are many things that our space agency could be doing to prepare for the adventures of the new millennium, and we think that NASA should be doing them now, instead of just continuing to fly the shuttle on trip after trip to nowhere.

But, of all the things that the space program needs, the one that it needs most is *imagination*, on the part of political leaders, in the administrators and managers of NASA, and by the captains of industry that run the giant aerospace companies. All of these people are presently acting like the commissioner of patents of a hundred years ago who resigned his office on the grounds that "everything has already been invented." NASA guards its tiny hoard of lunar rocks as if it were impossible to ever go get some more. The glamour, romance, and daring that characterized the days of Camelot in the 1960s are gone. We need to revitalize our spirit of adventure if wc are ever to grasp the promise that the future of space holds for us. Perhaps the announcement made in August 1996 that a meteorite from Mars may contain signs of extinct extraterrestrial life forms will be the trigger that reignites our excitement about the next major goals for humans in space.

## Note

1. Donald Cox and Dandridge Cole, *Islands in Space: The Challenge of the Planetoids* (Radnor, Penn.: Chilton, 1964), p. 63.

# 14

# Onward to the
# Main Belt Asteroids

I want to help build colonies in space when I grow up.
—Chelsea Clinton (Age 12), August, 1992

We noted in the last chapter that the easiest asteroids to explore are the ones that come close to Earth. Even though these deserve our first attention, the majority of asteroids exist in the "main belt" between Mars and Jupiter. Sooner or later we will want to go there and explore for the minerals described earlier. This will involve a trip away from Earth of several years' duration, a much longer-term proposition than that described in the last chapter.

## Mission Trajectories

The first modern thinker to plot trajectories for space flights to neighboring planets was a German mathematician, Walter Hohmann, who published a book on the subject in 1928 called *Die Moglichkeit Der Weltraumfahrt* (The Possibility of Spaceflight).* The most economical rocket

---

*(Munich: Oldenbourg Publishers, 1925).

trajectories that he computed to take one from the Earth to any of the planets are usually referred to as "Hohmann Orbits." Hohmann made it clear that the easiest way to reach another planet in our solar system—like Mars—was to first make sure that the spaceship escaped from the Earth in such a direction that it would eventually come close to the orbit of Mars. Ideally, the transfer ellipse should have an aphelion reaching just to the orbit of Mars, with a perihelion at Earth's orbit. The converse process would hold true if one were to fly to the inner planets of Venus or Mercury, with their transfer orbits at aphelion touching that of the Earth, and the spaceship at perihelion just reaching the orbits of those planets.

It has been shown, mathematically, that the techniques proposed by Professor Hohmann are the most economical way possible to move from orbit to orbit, and his technique for moving a spaceship from orbit to orbit is still known as the Hohmann Transfer. Until we perfect much more efficient rockets for moving about the solar system, we will be forced to rely upon this method for getting about. Unfortunately, these trips require considerable time for getting to Mars and beyond. Depending upon our destination, we will require one to two years to get to the asteroids in the main belt. It will also take that long to get back, so the total trip time will be anywhere from two to four years.

The main consequence of these longer trip times is that we will need a lot more life support supplies to sustain our crew. They will also require more living space than will suffice for a trip of a month. Further, this crew will need to be fully self-sustaining for a trip of that duration. This suggests that we will need to have a medical clinic, perhaps a psychologist, and other skills that we can neglect on trips of a month or so. A crew of couples may be needed to assure psychological health for such a long flight.

Another big difference between a trip to the asteroid belt and to a COTE near Earth is that we will need much more propulsion to approach any of the asteroids in the main belt. By the time we arrive in the belt, our velocity will have slowed so much that we will be traveling much slower than the asteroids in the belt. If we wish to stay there and visit some of these asteroids, we will have to increase our velocity. Then, to return home, we will have to reverse the process and reduce our speed so that we will fall back toward the sun and Earth.

## Traveling Asteroid Hotels

In chapter 13 we remarked that we could have stayed on Toutatis, if we wished, and ridden it all the way to the main belt, which just may be an excellent idea. One of the big concerns about a long journey in the solar system is protection from radiation storms. By burrowing several feet into Toutatis, we can have a storm shelter far better than any that we could bring from Earth. And since we have two years to ride Toutatis before we reach our destination, we could spend our time continuing to tunnel into Toutatis to build almost resort hotel-quality accommodations for ourselves, and for the next group of passengers who wish to visit the main belts. In fact, we can probably build tunnels to house huge gardens that will supply us with both food and air. Thus, Toutatis can become a "traveling hotel" for visitors to the asteroid belt. Note that the mass of Toutatis is expected to be something like 50 billion tons. This is a vast ore body, if only we knew of what it consists.

Unfortunately, if we wish to ride the same hotel back to Earth, our stay in the asteroid belt must be quite short, since this hotel never stops to wait for the trip back. The ideal solution would be to have a series of such traveling asteroid stations spaced so that there will be one leaving "soon" from any part of the asteroid belt. We do not as yet know of enough suitable asteroids to set up this transportation system, but we are discovering more all the time, and if the recommendations we (and others) have made to search more vigorously for Earth-approaching asteroids is carried out, the odds are very good that we shall discover enough suitable ones for our purpose.

## Is the Voyage of Toutatis Valuable?

Pending the development of a series of asteroid hotels, is there any reason to build the first one? Can we do any useful exploring from just one such asteroid? The answer is "of course we can." Even without getting off Hotel Toutatis, we can discover new things. We will come close, in comparative safety, to a great many main belt asteroids. Of course, we must be careful to define what we mean by "close." From the Earth, the main belt asteroids are anywhere from 300 million to 750 million kilometers away. At these distances, they are just tiny points of light to even the largest telescopes. From Hotel Toutatis we may expect to get within a million kilometers of many asteroids. At that range we will be able to see their size and their rotation, and if they have companion moons, such as Ida and Dactyl, we can estimate their mass and density. We can carry out spectrographic analysis of their reflected light, and perhaps even carry lasers powerful enough to vaporize bits of them for even more detailed spectrographic analysis of their surface materials. There is a wealth of scientific and prospecting data that we can obtain from such a mission. And prospecting is the proper word, since we do intend to return and mine these asteroids for minerals needed on Earth.

We may also consider the prospect of "scout ships" that depart from our asteroid home to visit promising targets and return to Toutatis to complete our four-year journey of discovery. We should note, in passing, that leaving our asteroid hotel exactly at apogee to go exploring is not that important; if we leave a month or two before and return a month or two after the precise time, it will cost little extra in propellant to do so. The trip plan is quite similar to the one we described for visiting Toutatis from Earth, except that now we will leave Toutatis to visit some site of interest and then return to Toutatis for our return to Earth.

Moreover, these same scout ships can be utilized to make close fly-bys of any interesting objects we encounter on our journey into the heart of asteroid country. Such "side trips" can add a great deal of scientific and prospecting value to our journey. It must be noted that each side trip will require a great deal of Delta-V. We will need at least 12 or 13 km/sec of Delta-V to accomplish each side trip, even in the last few months of our approach to Toutatis's aphelion. This is even more than we need in a SSTO space launcher. It may be easier to achieve increased Delta-V in a space-based spaceship, even if it uses chemical rockets as present spacecraft do, because we can use much smaller engines, and the fuel tanks and structures can be lighter, since they do not need to withstand the high accelerations needed to climb from Earth through a thick and turbulent atmosphere. Nevertheless, to provide such a large Delta-V we may have to resort to stage-and-a-half or two-stage vehicles.*

This raises the question of Toutatis's composition. Is it a carbona-

---

*We need to clarify what we mean by a "stage and a half" vehicle in this context. We are trying to describe a vehicle which keeps its propulsion and guidance systems intact throughout the flight, but which uses an extra, disposable, propellant tank, as the space shuttle does. This will permit the mass fraction to be smaller than if all of the propellant had to be stored in one tank.

ceous chondrite type of asteroid, an extinct comet, or an iron meteorite? The answer to this question will influence greatly how we exploit it in our exploration and colonization of the asteroid belt. If it is an extinct comet, then there is a very good chance that we will find vast quantities of water in its interior. We can break this down into rocket propellants, so that we can undertake many side trips on each voyage into the belt, provided that we take along, or can find on Toutatis, the materials that we need to make the drop tanks for our "scout ship."

If Toutatis is iron, we may not be able to use it as effectively to help in our exploration of the asteroid belt, but it will be an exceedingly valuable source of raw materials for constructing large projects in the Earth–moon system. It can even be a source of iron for Earth industry, since a very modest Delta-V shove can put the raw material onto a course where it can be steered to an Earth landing, slowed by atmospheric drag.

If Toutatis is carbonaceous in nature, we can expect to find enough water and aluminum to supply many scout ships, along with many other minerals of value to our intrepid band of explorers. Given the uncertainties, we will want to send a survey party to Toutatis as soon as we reasonably can. We believe that it would be possible to do this by the time the asteroid passes close to Earth in 2004.

## An Orbit Adjustment for Toutatis

We need to point out that the orbit of Toutatis is not exactly ideal for our purposes of mining or as an "asteroid hotel." Toutatis's orbit is about 3.99 Earth years. This means that, while it is making several close approaches in the near future, it is also drifting farther away from Earth over time. In a couple of decades, it will not be possible to travel on Toutatis in the manner we have been describing here.* But it it is entirely possible that

*This is a matter of having commensurable orbits. If an asteroid has an orbital period of just two years, it will come close to Earth every other trip around the sun, or every two years. However, if its period is 2.5 years, it will have to make two trips around the sun (while the Earth makes five) before they are again close, at the end of five years. If the periods of asteroid and Earth are nearly the same, as in the case of Toutatis, they will be close together for several trips around the sun, they will gradually drift apart, and then there will be a long interval (hundreds of years) before they are again close to each other.

we can correct this in the near future. The speed of Toutatis at perihelion (its closest approach to the sun) is fifteen miles per hour slower that we would like. If we could somehow add this much velocity to it, the asteroid would have an orbit period of almost exactly four years and we could use this planetoid station for centuries into the future.

This will not be a simple matter of attaching a rocket motor to Toutatis and firing it for a few minutes. As we noted earlier, Toutatis is massive. It will take a lot of energy to give it the small velocity increment that we desire. Fortunately, the means to do this were studied by the Los Alamos "Interception Workshop" referred to previously. The paper detailing the proceedings of the workshop describes how a few hydrogen bomb blasts will be sufficient to move Toutatis enough to preserve its usefulness for our children and their grandchildren.[1] From the data in this paper, it appears that less than ten medium-size hydrogen bombs would be enough to adjust the orbit to our liking. The mass of these bombs is expected to be about twenty to thirty tons, or just a bit more that we need for our crewed ship for the first voyage. Thus, one or two more launches from Earth, as described in the last chapter, will give us the ability to place Toutatis at our service for centuries to come.

We propose that our space program, or better still a global space program such as the one we will describe later in this book, undertake the following Toutatis program: Send an unmanned probe to Toutatis in the year 2000 to get as much preliminary data as possible. This could be done by accelerating the *NEAR* and *NEARS* programs, preferably by adding earlier flights to their schedule, instead of displacing them from their present target to Toutatis. Then, in the year 2004, send a crewed expedition to survey Toutatis, and perhaps to make some preliminary preparations for the voyage to move it, which voyage could then occur in 2008. Several spaceships would make the trip. One or two would carry the crew required to supervise the placement and detonation of explosives and to measure the results of each blast. Another ship or two would carry the supplies needed, primarily the bombs, of which the Earth presently has more than an adequate supply. (They may as well be put to some peaceful use!) It is important to use several bombs rather than try to do it all with one, even though the latter would require less mass of bombs. We will want to make the adjustments to the orbit a little bit at a time. Set one off, measure the effect, and repeat as many times as necessary.

By the time Toutatis returns in 2012 (or even 2016) the radioactive effects will have diminished enough that we can again board the asteroid (at least in regions not directly affected by our bombs) and begin our voyages of exploration into the asteroid belt. If we do not do this on something like this schedule, Toutatis will drift away from Earth, and it may be several hundred years before it comes close enough to again do this maneuver.

We need to note that Toutatis is only one asteroid, and it only travels to one region of the sky. We would like to find yet more asteroids that approach Earth and which we can use in a similar fashion. They need to have a period that is a near multiple of the Earth's orbit so we will have frequent opportunities to use them, and we want them to have an aphelion close to that of Earth. There are a few candidates in the table of close approach asteroids in Appendix D1.

One such asteroid is 1994 CN2. It has an orbit period of 1.97 years, or close to two years. Its perihelion is 0.952 AU, and its inclination is only 1.43 degrees. This later point is important, because making a change in orbital plane is the most challenging maneuver in space flight due to the expenditure of fuel which would be necessary to shift to another plane. Hence, we need to find asteroids whose inclination is as near to zero and hence, the plane of the Earth, as possible. Unfortunately, 1994 CN2 is not scheduled for another close approach within the next twenty years.

Asteroid 1991 VK is yet another possibility. It has a period of 2.50 years, so we could use it every five years. Its inclination is 5.41 degrees, a bit more than desirable, but still within reasonable limits. It will make close approaches in 1997, 2002, 2007, 2012, and later years. It has the advantage of being a bit smaller than Toutatis, so that if we need to adjust its orbit a bit it will be easier. It also does not go as deep into the asteroid belt, but its five-year trip time is a serious disadvantage. We confidently expect the astronomers to find more asteroid candidates in the next decade or two. As we have said before, it is generally believed that no more than ten percent of all COTEs have yet been located and characterized.

Ultimately, if we are not satisfied with these natural cosmic spaceships, we may have to construct our own. This should not pose too large a problem for the space engineers of the mid-twenty-first century, especially since they will be able to use asteroid materials for the most part, and will not have to haul everything they need, especially heavy radiation shielding, up from the surface of Earth. We will probably start by selecting our asteroid materials from COTEs. Then we will detach the materials from their parent planetoids and direct them toward Earth. By using the Earth's gravity well and atmospheric drag, we will move this raw material into orbits very close to what we desire. (We may have to use some propulsion, bomb or rocket, to fine tune the orbit to our exact specifications.) Then, with the raw material and the construction workers all transiting the solar system together, we can proceed to build a science station that meets all of our requirements.

## Benefits of Asteroid Science Stations

We need to point out one thing at this point: Getting to the asteroid hotel, or science station, or whatever you choose to call it, does *not* reduce the velocity necessary to travel from Earth to the asteroid belts. We still need to match speeds with the outbound asteroid to get on board for the trip out, and we need to match speeds in the asteroid belt to visit the planetoids there. We have not changed that part of the problem. But we have done something else that is very useful. We have made it possible to take one set of life support equipment and complete living quarters which have very effective protection against space radiation and use them over and over again. We could build a ship that would take us to the asteroid belt without using any asteroids at all, but this would have to be a far larger, heavier, and more elaborate ship than the one we will need for the short hop from Earth to Hotel Toutatis or another asteroid. Hence, the use of an asteroid will make it possible (and attractive) for far more people to visit the asteroid belts in the near future.

In particular, scientists can be attracted to the voyage, since we will be able to provide computers, telescopes, and other laboratory facilities to keep them productively engaged during the trip. They can keep in touch with their planetbound colleagues via e-mail. Most scientists and engineers are already quite comfortable with this form of communication, which transmits via radio or satellite, and use it extensively to keep up with their colleagues. It should be noted that telephone or video conferencing will not be very useful over most of the trip, since the speed of light is much too slow to allow this form of "real time" communications.

This delay time will increase from seconds at the departure of Hotel

Toutatis from Earth orbit to as much as forty minutes or more when Toutatis is at is farthest distance from Earth. So voice mail will work, but telephone conversations will not. However, broadcasting in either direction is quite possible, since the viewer does not care that the events he is seeing took place many minutes ago. There could be a weekly TV feature show, "This Week on Toutatis," broadcast to the folks back home!

## Necessary New Technologies

As we move toward this stage of our spacefaring civilization we will start to need a lot of new technologies. As we noted in passing, leaving our orbiting science station will take a lot of Delta-V. If we must rely upon present technology, we will be quite handicapped in our quest for new knowledge and unlimited mineral resources. So, new high-performance propulsion systems will be one of our biggest needs. Let us consider three possible means of meeting our requirements.

But first, what are these requirements? Basically we need to be able to develop a lot of Delta-V without having to bring large masses of propellant (or of propellant-manufacturing machinery) from Earth. Conventional chemical rockets use two propellants (which may be either solids or liquids) which react with each other. This reaction, usually called combustion, generates a lot of heat, so that the products of combustion, which are mostly gases, are quite hot. These gases are then expanded through a rocket nozzle to a high velocity. Then, according to Newton's well known laws, an equal and opposite force, called "thrust" by rocket engineers, is generated. This force is directed to make the spaceship go where we want it to.

This process has several limitations. First, the energy of chemical reactions is limited, so that the amount of energy that can be released has a sharp upper bound. With the use of liquid hydrogen and liquid oxygen in our modern rockets, we are near the theoretical limit of what chemistry can do. Hydrogen and fluorine can do slightly better, but the exhaust products are so very toxic that no one in their right mind wants to use them on Earth. (And, since both are scarce in space, there is little incentive to use them there either.) Further, since the parts of a rocket must withstand the heat generated by combustion, current technology is near that limit also. In short, there is not much improvement that can be had in the *performance* of chemical rockets. However, there is a lot to be done to make today's rockets *cheaper* and much more *reliable*.

Thus, it can be seen that we must do one of several things to provide for more effective transportation among the asteroids. We must either find a better energy source to power our jet propulsion devices, or find convenient ways to use the raw materials we will find in the asteroid belt, or best of all, do both.

We already have one handy energy source developed and ready to apply to space propulsion: the nuclear reactor. We also have more than enough fuel to power all of the space travel we will want for decades in the warheads of nuclear bombs that the nations have now agreed to give up. We can use the plutonium from these weapons to power nuclear rockets for many years. This would be a true peace bonus, since we would be putting materials manufactured for war (at great cost) to a useful purpose far from our beautiful home planet. This would be the best way we can imagine to dispose of this very dangerous material in a productive manner!

## Nuclear Steam Propulsion

There are two means by which we can utilize the energy from a nuclear reactor to power jet propulsion engines. One is the thermodynamic (heat engine) means of using the energy of nuclear reactors for space propulsion, the other is the conversion of nuclear to electric energy.

The thermodynamic approach runs a "working fluid," such as water, through the reactor. There it would be turned into steam and exhausted into space to provide the thrust we need to move about in the solar system. This is actually a very promising idea, provided only that we succeed, as the authors expect that we will, in finding large ore bodies containing ice in extinct comets posing as COTEs and threatening Earth.

This scheme has several advantages. It would utilize a well-known material which we can confidently expect to find in large quantities in the the asteroids, and which can very readily be extracted from its "ore body" and prepared for use in the propulsion system. A paper presented at the 1995 Space Manufacturing Conference at the Space Studies Institute in Princeton, New Jersey, discussed how to extract water from COTEs for various uses in the inner solar system.[2] The author, a geological and mining consultant, lists a number of near-term opportunities to rendezvous with COTEs and possibly extract water from them. He describes in detail the procedure for drilling into these bodies and extracting water. Either extinct comets or asteroids of carbonaceous chondrite composition will be useful in providing substantial amounts of water, and these are expected to be a majority of the COTEs that we have discovered.

We can get reasonable performance from "nuclear steam" rockets. The "propellant tanks" can be very light. If we keep the water in the form of ice and use low accelerations (a fraction of an Earth gee) we really need no container at all, just a means of feeding the ice into the propulsion system before it gets away from us. The specific impulse (the engineers' measure of rocket performance) will be modest, ranging from 100 to perhaps 400 seconds, depending upon how we wish to make the tradeoff between thrust and asteroid consumption. For a given Delta-V the lower specific impulse will give more thrust and thus be able to move a larger mass, but will use a lot more water from the asteroid we are using for "ore." If we need a lot of Delta-V (as for our scout ship from Toutatis) we will want to go for the highest specific impulse we can get. This will be limited by the reactor design.

Today's design concepts and materials will limit nuclear rocket performance to a specific impulse of 400 seconds or less (for a steam rocket). If we substitute hydrogen for the water, we can perhaps get a specific impulse of nearly 1,000 seconds. This would allow our Toutatis scout ship (with its 13 kilometers per second performance requirement) to be only 75 percent propellant. This will probably be possible in the twenty-first century, since by then we expect to be flying SSTO vehicles that are over 90 percent propellant. It is possible to imagine other nuclear rockets of much higher performance. One such idea postulates a "gas core" reactor, which will overcome the thermal limitations of the engineering materials now available. Who knows what the engineers of the twenty-first century will be able to build?

There is a major drawback to any use of reactor-powered spaceships. The radiation from the reactor is deadly to humans. To use such propulsion on a crewed vehicle means that a great deal of shielding must be provided. If we are moving a large asteroid with a very large nuclear rocket, we can burrow into the asteroid to get the shielding we need. However, for a small "scout" ship, this will not work, and we will need a massive radiation shield and/or a great distance between the reactor and the crew.

Current design concepts provide for both, and make for a very ungainly looking spaceship. Perhaps this problem can be solved, but just how is far from known at this time.

## Converting Nuclear to Electrical Energy

The other way in which nuclear power can be used for propulsion is to first convert it into electricity. Then there are a variety of electric propulsion schemes that are possible and which have very high performance. The simple rail gun mentioned earlier can be a propulsion device, able to use almost any conceivable asteroid material as "working fluid." Rock, iron, aluminum, peanut butter, anything would be suitable, since the only property we really require is mass that can be accelerated to a high velocity.

There are several problems with all of the electric propulsion schemes that have been worked on up to the present time, however. One of the biggest is the mass of the electric power system. Both the generators and the equipment required to transform the power into the form needed by the electric thrusters has been quite heavy in relation to the amount of thrust provided. Part of the reason for this is that there has been little incentive to spend much money and effort to make dramatic improvements. It seems clear that electric propulsion will always be limited to very modest thrust-to-mass ratios; it will never be useful for taking off from a massive planet such as Earth or Mars. Even once in orbit, the usefulness in achieving high velocities through electric propulsion will be limited.

The use of electric propulsion has, thus far, been limited to very specialized applications. For example, it has been used to provide "station keeping" propulsion for geosynchronous satellites. Since this can be done with very low thrust levels, it is a "natural" application for electric thrusters, but the reward for using it is rather small. In place of having a hundred pounds of propellant and ten pounds of rocket hardware such as might be necessary for chemically based propulsion, you can have ten pounds of propellant and forty pounds of rocket hardware. There is an advantage, but not a big enough one to make space engineers get very excited. That is especially true of their steely-eyed managers and their accountants. As a consequence, large sums of money have never been spent to develop electric propulsion. However, when the equation changes and we are discussing a hundred tons of propellant versus ten tons, with the machinery being only a small fraction of either amount, money will be invested in the technology, and we can expect to see significant improvements.

The other problem that has plagued electric propulsion to date has been a very limited lifetime for the thrusters. This is believed to be simply an engineering problem, and not a fundamental limitation. Our engineers have just not yet learned how to effectively contain and control the very high energy densities involved in electric thrusters, but there is no reason to expect that this will continue when enough money and attention are focused on the problem.

The final limitation for electric propulsion has been the energy source. Two sources are possible: nuclear and solar. If solar is used, then the collectors must be very large in order to obtain the megawatts of power required for moving crewed spaceships around the solar system. To date these have been much too heavy to provide really useful accelerations. From fundamental principles, it seems likely that this will always be the case. Near Earth, we require at least a tenth of a gee to move our spaceships about in Earth's deep gravity well. We can obtain a kilowatt of sunlight energy for every square meter of collector (if we can use all that we collect, which is still beyond our technology), but even at that density, the collector has too much mass. In the asteroid belt, we can get by with a lower acceleration since we are not in a intense gravity field, but the solar intensity is also ten times less there. The other energy source, nuclear, has never been developed in the many-megawatt-size package needed for space propulsion. Further, as we noted before, nuclear energy of any sort requires a massive radiation shield to protect the crew of any vessel it propels. So it, too, has been limited to very low accelerations in the past, and prospects for any great improvements are not promising.

One of the ways around this dilemma is to separate the power source from the spaceship being propelled. This can be done by "beaming" the energy from a power source to the spaceship under way. Most of our readers will probably have heard of the proposal to build a huge (miles by miles) solar power station in high Earth orbit and then to send the energy down to collectors on the ground by very powerful microwave radio beams. The scheme will work, at least in principle, although there are serious questions about its economics and its potential effect upon the upper atmosphere. However, the principle has been established, and enough testing has been done to determine the engineering design factors needed to proceed with a full-scale design.

This principle can be used to provide the energy for an electric propulsion system for a spaceship, either near Earth or deep in the asteroid belt. We could use a solar power satellite to escape from Earth, and a nuclear power station on convenient asteroids to power our ships there. The reason this can be made to work is that we are now not limited to the energy density of sunlight. We can transmit a hundred times the energy of a beam of sunlight to our spaceship, allowing the collector to be much smaller. Further, the conversion efficiency of this radio energy to electric power can be 90 to 95 percent. The best we can do with sunlight, so far, is more like 10 to 30 percent. Hence our power collector can be made very light. If we can eliminate most of the power conversion mass, and there is no apparent reason why we cannot, then we may be able to have high-performance electric propulsion systems with enough thrust to accomplish our space missions.

## Elbow Grease on the Planetoids

There are several other new technologies that we will need, but we will close this section by describing just one more. We will need tools and techniques to work on our little planetoids in the almost complete absence of gravity. We talked earlier about mining/tunneling into Toutatis to build our radiation-proof habitats. If you have ever looked at the construction machinery used on Earth for doing this kind of work, you will immediately see that the very principles will just not work on a planetoid. Earth-based construction machinery is huge, and it depends upon the force of Earth's gravity to hold it in place so that it can move the rock and dirt involved in digging a tunnel. Since Toutatis has essentially no gravity, we will have to use some other principles to dig tunnels. Perhaps we can drive the equivalent of tent stakes into our asteroid to hold our machinery down. However, if we are working on an extinct comet, we expect that it will be too fragile for that to work, and we will have to devise some other scheme. Dynamite? Lasers? Pick axes? Ooops! A pick axe will not work either. When we try to bring it down, Old Man Newton will have the last laugh, and as the axe descends, we will go up!

It is clear from even this brief description that the managers of our space program have given little serious thought to the kinds of things that they should be doing to prepare for the space exploration that lies ahead in the twenty-first century. It is time for some imaginative thinking to begin in the halls of NASA, our space companies, and the space agencies of other countries.

## Notes

1. John L. Remo and P. M. Sforza, "Near Earth Object (NEO) Orbit Management by Explosive Impulse Thrusters," in *Proceedings of the Near-Earth-Object Interception Workshop,* G. H. Canavan, J. C. Solen, and J. D. G. Rather, eds. (Washington, D.C.: GPO, 1993), p. 194.

2. David L. Kuck, "Exploitation of Space Oases," in *Space Manufacturing 10: Pathways to the High Frontier,* Barbara Saughnan, ed. (Washington, D.C.: American Institute of Aeronautics and Astronautics, 1995), pp 136–56.

# 15

# Why Colonize the Asteroids?

The Earth is the cradle of the mind. . . . Mankind will not remain on Earth forever, but in its quest for light and space, will at first timidly penetrate beyond the confines of the atmosphere, and later in the search for heat and light will conquer for itself all the space near the sun.
—Konstantin E. Tsiolkovsky (1857–1935),
the "Father of Spaceflight"

What is the destiny of man? What is the purpose of human life? If no one knows the answers to these questions now, can we ever hope to find them? We have searched for the answers on Earth for thousands of years without much success. Possibly we can find them out among the stars.

Our world will not live forever, and long before it—or the sun—dies, such violent changes will take place on the surface of the Earth that life will no longer be possible. Will humanity die with the Earth? Perhaps these questions appear completely theoretical and of no real practical importance to anyone. And yet, a lecture by an astronomer at a planetarium on the subject "The End of the Earth" will always draw a big audience.

This should not really be surprising since it deals with some of the most fundamental of human motivations. Why do we want to discover new knowledge? Why do we want to go on living? Why do we want children?

211

We do not know why. Perhaps the same fascination with the unknown that lured the early explorers to America and then urged them to conquer the West now motivates us to probe ever deeper into space. We find that most—but not all—people have a strong interest in the future of the human race and the destiny of our species. It matters to us whether we *Homo sapiens* are destined to die with the Earth just as it matters what happens to our grandchildren. Perhaps we are all linked by undiscovered bonds. Perhaps that is why it hurts us when we harm someone else and why the whole world may grieve at the loss of one popular political or religious leader.

In any case, the question "Why do we need colonies in space?" may not need a detailed answer. It may need no more answer than the young pioneer in his one-room log cabin needed when his wife told him that it was time to start building a new room. It is simply our nature to colonize new islands, new continents, and new planets. Perhaps the really sensible question is not "Why should we colonize space?" but "When should we start?"

## Population Growth and Earth Exodus

Recently it has been fashionable for intellectuals to be very concerned about the population explosion, to campaign for free distribution of birth control information, and so on. As everyone knows, there is a very serious overpopulation problem in some of the underdeveloped areas of the world and present prospects for improvement of the situation are not encouraging. People now talk about the "carrying capacity" of the planet, in the same way that range managers talk about the "carrying capacity" of grazing land for cattle.

However, the population question in the advanced technological nations is not the same as in the less developed nations, and it is not at all certain that we will be confronted—in this century—with a serious problem comparable to that of the less developed nations, regardless of how much our population increases. The population of the United States is growing at a rate such that it is expected to double in about forty years. However, our gross domestic product is expected to double in twenty years and our technical knowledge even faster. This suggests that we may be able to maintain our standard of living, *provided* that we find the natural resources we need. There is reason to believe that undeveloped countries may follow the same pattern, once they "catch up" in terms of education and standard of living.

Nevertheless, it may be that the most important reason for colonies in space will be to provide for continuous growth of the race when growth is no longer possible on the Earth. Long before growth comes to a complete halt on the Earth, it will have to slow down. And even before it slows down, population pressure will bring many restrictions and limitations which will eventually become intolerable to the independent pioneering spirits among us. Those who cannot find a place in the increasingly more rigid societies of Earth will look for new lands where there are no rules, or where they can make the rules as they want them. They will go out seeking knowledge, wealth, power, or uncrowded places, but mostly they will go, as pioneers have always gone in the past, seeking new dimensions of freedom.

The nations of the world have the capability to control their population growth. Some nations, such as Japan and China, have already taken rather drastic steps such as legislating family size and have succeeded in halting formerly explosive population growth. The capability to control population presents each nation with a choice. The nations can take steps to slow down or halt their growth (as Japan and China have done); let nature take its course; or try to encourage more rapid growth, ultimately resulting in disastrous consequences for us all. However, if a nation chooses to limit growth, that nation may decline in its fraction of the global population and thus in relative global influence and power. Its culture might be lost among the teeming masses. As will be discussed shortly, the possibility that all nations will act to limit growth and thus maintain the ethnic percentages across the globe does not seem likely. Therefore, we do not support this alternative.

How can the advanced technological nations resolve this dilemma? Must they choose either the course of relative decline through population limitation or the crowding and loss of individual freedom which will result from overpopulation? The other alternatives appear to be war, international agreements to limit population, and spaceflight.

Of these three, the first is just unthinkable. In recent years, humanity has made yeoman efforts to outlaw war. Success in this endeavor is limited, to be sure, but it remains an ideal to which people of good will everywhere are dedicated. The second alternative, international agreements on population control, is even less likely than successfully outlawing war, since there is widespread religious and ethnic opposition to limiting births. Only the third alternative, space colonization, appears both desirable and attainable. Only by colonizing space can the advanced nations continue to hold their present positions in global power and prestige without serious overcrowding and the resultant increase in regimentation and government control.

## Global Catastrophes

There are one thousand and one ways in which the great adventure of the human race on the Earth could be terminated, but the one generally selected as the most probable is full-scale nuclear war.

Would nuclear war actually end all human life on the Earth? No one really knows. Probably not immediately. But the survivors of the war might be so weakened by radiation sickness and disease that they could not long continue the struggle to survive in a devastated world. Ever since the explosion of the first atomic bomb, people have worked to somehow balance or alleviate this terrible new threat to our security. These efforts—including the formation of the United Nations—have been partially successful, and more progress toward stability and world law may be possible in the future. However, there is no sure way to completely remove the threat to the whole human race—*as long as we all stay on the Earth.*

Of course, there are many other ways in which all human life could be destroyed. Some are so-called natural catastrophes and others might be the result of tinkering with the delicate balance of nature. Among life-destroying natural catastrophes we will mention the possible collision of the Earth with a large planetoid or comet, a decrease or change in the quality of the sun's energy output, increase in the sun's output (including explosion of the sun), violent changes in the weather, rearrangement of the Earth's crust, serious upsets in the Earth's ecology, rise of a mutant virus of unprecedented virulence, and so on.

Possible manufactured catastrophes include many on the above list with the difference that they might be brought about through our own meddling rather than by natural processes. One example is global warming. If this theory is true and we continue our present ways, we will affect the natural environment in a negative way. In addition, we must consider the possibility that some insecticides, medicines, detergents, food processing chemicals, etc., will come into worldwide use before it is discovered to have a delayed lethal effect. Suppose, for example, that the product consumed was even more commonly used than tobacco, particularly among the young; that the lethal effect was produced after about ten years; and that it was even more deadly than tobacco. We can imagine some kind of soft drink or confection becoming universally popular among the young of all countries and races. Then suppose that some flavoring or processing chemical had a delayed reaction of perhaps ten years on one of the body's vital organs. Perhaps some area of the brain would be damaged to the extent that the children would never grow up, or perhaps their reproductive capability would not develop.

Of course, the probability of a worldwide catastrophe resulting from any one particular kind of natural or manufactured event seems so low as to be of no real consequence. However, if we multiply this vanishingly small probability of a catastrophe occurring in, say, the next ten years by the number of possible occurrences, and extend the time period into the indefinite future, then the picture changes drastically. The vanishingly low probability rises to something very close to certainty. In other words, a major race-destroying catastrophe is bound to occur if we wait long enough!

## Planetoid Mobility

The potential mobility of an inhabited planetoid is, in itself, a reason for colonization since the planetoid can be moved to exactly the orbit desired. This could be an orbit in the planetoid belt, an orbit around the Earth, or perhaps an orbit closer to the sun to make use of free solar power as an alternative to nuclear power. Planetoid colonies might someday move from place to place in the solar system looking for the best spot to apply their resources and human skills in the constantly growing and changing space economy.

In the more distant future we can also imagine some planetoids being moved out away from the densely populated inner solar system to the vast emptiness beyond Jupiter. With a plentiful supply of nuclear power these colonized islands could drift gradually away from the sun, visiting the satellites of the outer planets—perhaps planting new colonies—and eventually passing beyond Neptune and Pluto and setting sail on the black seas of interstellar space.

# 16

# Inside-Out Worlds

That inverted bowl they call the sky,
Whereunder crawling coop'd we live and die.
—"The Rubaiyat of Omar Khayyam"

As we suggested earlier, the occupation of the asteroids will probably take place in phases, beginning with the landing of the first explorers on Toutatis, or some similar COTE, for the long voyage to the asteroid belts.

## Beachhead Bases

The first protective shelters on the planetoid beachheads to serve as spartan space-mining camps will most probably be improvised from small shelters and empty propellant tanks from the upper stages of transport rockets. These small two-to-four person shelters when joined together will suffice for a time as the first primitive space environment living quarters to be set up on the surface of the planetoids. After assembling several of these dual-purpose modules (which would not be needed for the return flight to Earth), following an expandable base concept, plans could then be undertaken to enlarge the base into a permanent installation.

217

Such hardware as the life-support system, a nuclear power plant, and regenerative fuel system for powering the surface vehicles used in conjunction with the larger planetoid bases will next be installed, using portable packaged systems flown from Earth. These systems will be interconnected with the modules by cable and pipeline. The power plants and modules will most probably be buried beneath the surface, with the help of small mechanical moles, to protect the astronaut volunteers from the lethal effects of solar radiation and the bombardment of any stray meteorites.

The experience gained in building subsurface installations in the Antarctic, which are covered with ice and snow, should be of inestimable value in helping our space explorers to establish their first under-surface extraterrestrial bases. (The first bases established in space might be on the moon. There is currently more interest in a return to the moon than in visits to the planetoids. However, the planetoids are easier to visit, and with the renewed recognition of their threat to our future, they may well be the first extraterrestrial habitats. But if the moon *is* first, the experience gained there will be directly applicable to the development of planetoid bases since conditions will be quite similar.)

The first quartet of astronauts to land on a minor planet might very well consist of a mixed crew, consisting of two space pilot-navigators and two scientists, with one preferably being a specialist in geological sciences and the other in space physics. While one twin team remains inside the module beachhead for safety and communications purposes, the other team consisting of a pilot and scientist can range out from the base headquarters for a few hours to a few days.

These roving expeditions would explore the planetoid's potential with the aid of miniaturized Geiger counters and other sophisticated mineral detection devices. Photographs as well as rock samples will be taken back to the command post for eventual analysis and relay of data back to Earth by radio or as actual cargo on the return flight.

Once the beachheads on the islands have been secured with portable igloo-type space homes flown up from the Earth or by utilizing the capsules themselves as a temporary home, then plans can be undertaken for the construction of more sophisticated dwellings. From the knowledge and experience that we will achieve a few years hence in the establishment of permanent bases on the moon, we will be able to adapt many valuable lessons for application on the planetoids.

But unlike the dusty, and probably porous, surface of our lunar satellite, the typical planetoid composed of rock and metal will offer us different challenging opportunities for the construction of permanent cosmic cities that will undoubtedly follow in the wake of the outposts established by the early astronaut explorers. Creating bases on the planetoids for reasons such as science, the military, "filling stations," mining, etc., also provides the impetus for colonies, since a base will become cheaper to maintain as it becomes more self-supporting. Also, the working staff of the base will be more content to stay on the base for long periods without return to Earth if their spouses and families can be with them.

## Journey to the Center of the Planetoid

The possibility of hollowing out planetoids for space colonies was discussed some years ago by L. R. Shepherd, I. M. Levitt, and other scientists and science fiction writers. It is believed that the first discussion of this possibility was by the late English scientist Dr. J. D. Bernal in his book *The World, the Flesh and the Devil.** In the early 1950s, Dr. L. R. Shepherd, a former chairman of the British Interplanetary Society, expanded on an earlier idea proposed by Bernal for gouging out the insides of a planetoid. Shepherd envisioned the crews living inside the shell of the captured minor planet, which could be spun to provide a simulated gravity condition; i.e., to eliminate weightlessness. Of course, at that time no one knew of the many COTEs which have since been discovered.

The new half-artificial, half-natural space station would make an ideal home for the astrocolonists who could live on the excavated floors, or even on the inner walls of the hollowed-out planetoid like the ancient cliff-dwellers of southwestern Colorado. This latter practical possibility, which is really borrowing an idea from the ancient Mesa-Verde Indians, was not envisioned by Shepherd when he formulated his hypothesis.

But where will the astronauts deposit the excavated chunks of rock and metal as they burrow deeper into the interior of the minor planets? At first, they can pile the debris from the scooped-out insides of the selected planetoid on the surface like an anthill. This material can be refined and

---

*(New York: Dutton, 1929).

utilized as fuel in rocket engines returning to Earth or moving out into deeper space, or for life support and building materials. Hopefully, little will be wasted.

A hollowed-out planetoid would provide living space and would also protect the inhabitants from the danger of meteorite impacts. An artificial space station constructed by humans, on the other hand, could not economically have walls as thick as a planetoid space station to shield the astronauts from penetrations of high-speed celestial objects and deadly cosmic radiation.

A typical cannibalized planetoid would progress from a single small compartment to a series of adjoining compartments—each serving a different, but related, function. The crews who will staff these slowly spinning giant space stations will move around with their heads pointed toward the center of cavernlike ships. Dr. I. M. Levitt, the former director of the Franklin Institute's Fels Planetarium, has envisioned these captured and cannibalized planetoids as potential "celestial Noah's Arks" because of their functional resemblance to the biblical ark. They may serve as the homes for the first large colonies in space and these celestial arks can theoretically be made to maintain miniature daughter civilizations of the Earth.

Air locks can be bored in the sides of the minor planets selected for exploitation and capture to serve as entrances to the interior. They would serve in a similar capacity to the small outer tunnel separating an Eskimo's igloo from the Arctic elements, or to a submarine diver's decompression chamber. They could be protected with a tough outer shell which would be strong enough to withstand the impacts of the constant hazards in space: micrometeorites and cosmic dust. These air lock portals would provide entry into the interior of the asteroid, where the walls would be so thick that almost no impact could threaten the crew.

Gradually, over the course of many years, a major fraction of the material of the planetoid would be processed and redistributed. This could lead to a great expansion in the habitable space in the planetoid as materials were carried outward and great empty spaces were created. Eventually, caverns extending for miles would take up the entire inside of the enlarged cosmic body and the colonists would have built themselves a new world.

While the first planetoid worlds will probably be hollowed out over a long period, the cosmic engineers and colonists of the more distant future might become impatient with such a slow process. When large-scale operations in space have become relatively commonplace, engineers will begin to think of ways of creating large, habitable, planetoid worlds in one process. They may consider taking an iron planetoid, melting it, blowing it up into a long cylindrical bubble, and then waiting for the iron to solidify and cool down to a temperature consistent with human habitation. Such a project would dwarf anything ever undertaken on the Earth. How could engineers possibly bring it about even with the techniques and machines of the future?

It would be foolish to try to specify exactly how an operation of this magnitude would be conducted perhaps fifty to seventy years from now, because of new scientific and engineering discoveries which will be made in the meantime. However, it may be useful to examine the project to see if it is compatible with what we know today about physics. John Campbell, the well-known editor of *Analog* science-fiction magazine, once suggested constructing large curved mirrors in space, and using concentrated solar heat for melting, cutting, and shaping planetoid materials. Dandridge Cole also suggested that giant solar mirrors could be used for forming the planetoid bubbles.[1]

Cole's concept for the world-makers would be the construction of a giant convex mirror which might measure several miles long. Holes would be bored down to the center line of the long axis and when completed, would be charged with tanks of water. The planetoid would then be set spinning slowly, such that the entire body would be bathed in the intense heat of the reflected and concentrated sunlight. Gradually the flying iron mountain would be heated to the melting point all over its surface, and slowly the heat would creep inward until almost the whole object was molten.

When the entire body was melted, the gravitational and cohesive forces would presumably pull it into a spherical shape. This would not be desirable, as we will see later, and would not occur if the engineers had done their job. The central axis of the planetoid would be the last part to melt. As long as this remained solid, the melting body would hold a cylindrical rather than spherical shape. Thus the engineers would design the water tanks to explode from internal steam pressure just as the central axis melted. Then the released expanding steam would blow up the planetoid into an iron balloon some ten miles in diameter and twenty miles long.

In principle Cole's scheme will work. However, it is useful to note that as the surface approaches the melting point of iron, it will radiate away vast amounts of heat that otherwise could go into speeding up the asteroid blowing process. At the melting point of iron, each square meter of planetoid would radiate away several hundred kilowatts of heat. It will take several thousand square meters of solar collector to offset this loss, so that it will take a very large collector and a long time (ten or twenty years) to melt an asteroid. One alternative is, of course, an even larger solar collector, but this is self-limiting, because of the large heat losses.

There are a couple of ways to reduce this problem. One way is to heat the asteroid from the inside. Using high power lasers, which were not available when Cole made his proposal, we can bore a small tunnel down the center of the asteroid, and send our beam of focused sunlight to the inside of the asteroid. This way, it will not lose heat so fast, until the very outside is just about ready to melt, and we can melt the planetoid more quickly. The solid outside will hold its shape until we are ready to inflate the iron balloon. In fact, we do not need to bring the outside quite up to the melting point; at near-melting temperatures the iron will flow quite readily as we inflate it.

Another, perhaps even more attractive, way is to enclose the asteroid we are working on inside a bubble blown from a much smaller asteroid. This bubble, which can be made from a very small amount of asteroid material, since it can be quite thin, surrounding the nearly molten asteroid, will reflect the heat back onto the asteroid which is the subject of our efforts instead of allowing it to escape to space. This will allow us either to use much smaller mirrors, or to accomplish our task much more quickly. It may be advantageous to use both these ideas in conjunction with one another to speed and simplify our work.

There is another important advantage to using a giant bubble to contain our "work zone." Unless we are very lucky, the target asteroid will not be pure iron; it is apt to contain other things which will boil off at a much lower temperature, such as water. By encasing our work-piece in a container, we can capture these materials and put them to our use.

Years later, when the hollow shell had solidified and cooled off, the construction crews would affix the giant mirror to one end and direct a beam of reflected light down the long axis of the cylinder to form a linear sun. The next jobs would be to set up plants for manufacturing air, water, soil, etc., for the new world. Raw materials would probably be obtained from a rocky planetoid which had been captured and pulled alongside.

Gravity for the new world could be supplied to the extent desired by spinning the planetoid about its long axis, like a rolling log. This would produce a centrifugal force which would seem to push objects against the outer wall. The advantage of a cylindrical as compared to a spherical shape should be obvious because of the larger surface area available with the desired artificial gravity.

Suppose for example, that the inhabitants selected to have a gravity equal to one-fifth of the normal surface gravity of the Earth (such that a 200-pound man would weigh only forty pounds). To produce this effect in a cylinder ten miles in diameter, it would be necessary to give the planetoid a spin rate of six revolutions per hour. At this rate, the circumference would be moving at 600 feet per second, but people inside would not be aware of this motion. They would only sense the apparent gravitational force caused by the rotation. In addition, people or objects in motion would experience a second apparent force called the "Coriolis force" which would appear to deflect the path of moving objects.

If someone standing on the inner wall of the spinning cylinder tried to throw a ball straight toward the center, they would find that it would be deflected away from a straight line on a curved path in the direction of rotation. If we say that the cylinder is turning toward the east, then balls thrown straight upward would be deflected toward the east. On the other hand, a ball dropped from a height would not fall straight down but would be deflected toward the west.

To a person raised inside this strange world who was never told of its spinning motion, both the centrifugal and Coriolis forces would be as mysterious and unexplainable as gravity is to us. However, if the person happened to be a physicist, he could eventually figure out that his world was turning, and even calculate the spin rate. Some scientists believe that our own gravity is just the result of some strange curved motion of the universe that we will probably never be able to visualize.

We have discovered that there are some human limitations on using spin to provide artificial gravity. If we spin our asteroid too fast, our middle ear will tell us that we are turning and upset our equilibrium. Although we have not had any centrifuges with a ten-mile diameter to do tests with, we are reasonably confident that this is large enough that even the most susceptible people will not experience discomfort.

The gravity of the tiny inside-out world would decrease from one-fifth Earth normal at the cylindrical surface to zero at the central axis and at the poles at the ends of the central axis. Thus a 200-pound man who traveled from the equator to one of the poles would find that his weight dropped from forty pounds to zero during the process.

The inside-out world would be constructed with a slight built-in equatorial bulge so that the equator is always downhill. Thus, water from the melting pole caps would flow downhill in rivers and streams to the central equatorial sea. The low temperatures of the poles would be maintained by great disk-shaped shades near the ends of the central sun which would keep the poles in perpetual darkness. Also, thermal radiators would be mounted outside the planetoid at both ends to reject heat and balance the continual influx of heat from the mirror. Water vapor which evaporated from the rivers and the sea would tend to condense and freeze at the poles. At the mass of ice and snow piled up, it would tend to slide downhill into the unshaded sunlight, where it would melt and run back to the sea.

The central beam of sunlight itself would probably shine down through a transparent hollow tube. The tube would be filled with a gas of the required composition and density to produce the desired scattered and fluorescent light and heat. The "sun" could be turned off at intervals by blocking the mirror if periods of nighttime were desired.

Soil for the inside surface would be made from pulverized rock and the decayed remains of several cover crops. A parklike or more natural appearing random countryside would be landscaped to suit the desires of the colonists. The ten-mile-diameter cylinder world would have an inside surface area of 628 square miles, over half the size of the state of Rhode Island. Uncrowded Rhode Island has a population of 800,000 people. Our inside-out world could very comfortably hold perhaps 100,000 people. People of the modern world do not dream of paradise as much as did their ancestors of even the last century. But there may still be some who dream of a physical paradise and for them, and for all the dreamers of the past, the hollow planetoid could be a possible escape outlet.

Within the inside-out world, all the physical conditions of the environment would be completely under the control of the inhabitants. The temperature, air pressure, humidity, oxygen content, pollen content, rainfall, snowfall, day-night cycle, etc. would all be under complete control. The quality and quantity of food produced, the number and type of trees, shrubs, flowers, and streams; the wildlife, birds, fish, etc. would all be as desired. There would be no harmful microorganisms or insects, only those needed to maintain the balanced ecology. There would no pollution of the air or water since all biological and chemical wastes would require reprocessing in any case and there would be no point in dumping them into the gas and liquid reservoirs.

Very little physical labor would be required since all routine tasks would be done with automated machinery. This would apply also to routine clerical and mental work, such that drudgery would be almost completely eliminated and inhabitants would be able to devote their time to nonroutine, creative tasks. The control over the environment of this Utopia would be so substantial that the inhabitants could even control the force of gravity for their world. If they wanted a full Earth gravity, they could have it. If they wanted their weights to be one-fifth or one tenth of Earth normal, they could arrange that also. Or they could cut the force of gravity down one hundred or one thousand times below Earth normal, and could even change it back and forth from time to time.

Reportedly, one of the commonest types of sleeping dreams is that of flying freely through the air using only our own arms for propulsion and support. This dream is probably as old as humanity and perhaps even older, for even the pre-*Homo sapiens* probably dreamed. Our present ability to fly through the air with high-powered mechanical contrivances does not satisfy this universal desire for freedom from the force of gravity. A person who believed in prophetic dreams might argue that we have actually foreseen the future existence of these low-gravity hollow worlds in our dreams of unpowered and almost gravity-less flight, because we know of no other way in which these dreams could ever be satisfied.

*Imagine for a moment that you live in this hollow world in a little house in the country and that the spin rate has been set to produce one-tenth of Earth normal gravity. You decide to go for a short flying trip and a breath of country air. You put on a flying suit with paddle-like wings attached to the sleeves and extending about two feet beyond your hands, and light cloth sail extending between arms and legs. You give a slight push, glide across the house and out through the front door, folding your wings momentarily to make your exit. Outside you land on your front lawn and get set for your take-off spring. Since you are fairly athletic and have strong legs, you shoot straight up with wings closed to a height of thirty feet. (On Earth you could raise the center of gravity of your body three feet straight up, which is rather good.)*

*At thirty feet, high over the top of your house, you come to a stop and maintain your altitude by flapping your wings. This is not difficult since you have a mass of 160 pounds and a weight of only sixteen pounds, no more than a small turkey. Thus, you need support only eight pounds on each arm, which is no strain even for someone who is not in top physical condition. You look around briefly at the rolling hills, the green forests, and the open meadows of your homeland and then drop into a steep glide down the valley. The dive brings you speed which you lose again as you pull up to recapture your altitude, but this time you flap your wings vigorously and ascend to some seventy feet before resting in a longer but slower glide.*

*In passing a stream at the lower end of the valley, you spot a trout just breaking water and throw yourself into a steep dive. You make a playful try at catching the fish, who has risen over ten feet out of the water, but you misjudge the correction for the Coriolis force and pass to one side. But now you are heading for the stream at high speed and with considerable momentum. Your small wings are not sufficient for a sharp turn and you end up ingloriously in the stream, producing a splash of water which rises thirty or forty feet.*

*While this inauspicious beginning of your outing dampens your clothing, it does not dampen your spirit and you spring into the air with new energy. You start out to set your own personal altitude record. Flying straight up is really no harder for an experienced flyer than brisk walking on Earth, and after a half hour of steady flying you have risen to a height of some two and a half miles. Half-way to the central sun of your world you find the flying much easier since your weight has been reduced to only eight pounds, but it is also becoming uncomfortably warm. Remembering the experience of Icarus, who flew too close to the sun and melted his wings, you decide to forego any further conquest of altitude and settle for lazy contemplation of the beautiful scenery on your gentle glide back to the ground.*

*From this height you can get a much better impression of your cylindrical world than from down on the ground. Through the clear air you can easily see the pole cap about five miles to the north and the equatorial sea almost an equal distance to the south. In fact, you can get a view of the entire world-circling equatorial sea and even make out the boats on the far side sailing upside down across your sky. The up-curving and encircling horizon is not disturbing to a native of the inside-out world, and even visitors from Earth quickly lose their concern. They soon real-*

*ize that the ships, vehicles, people, houses, etc., that they can clearly see through binoculars above them on the opposite side of the world are not going to fall on them. They gradually adjust to the upcurving horizon and begin to enjoy the greatly expanded view of the scenery that this affords. Any small hill in the hollow world gives the sightseer a broader panorama of forests, hills, meadows, villages, lakes, etc., than even a view from a high mountain on the Earth.*

*So you drift slowly back toward your home with your clothing well dried by your close approach to the sun and a great feeling of peace and contentment with your happy, worry-free life in this utopian world. Your exercise has stimulated your appetite and you speculate idly on what selections you will make for the evening meal. After supper, of course, you must attend the village meeting to settle that aggravating question of whether to have rain once or twice a week during the next three months.*

## But First, Baby Steps

We will want to start with much smaller objects than the Rhode Island-size object that Cole envisioned. Fortunately, we believe that there are plenty of smaller objects to practice on before we tackle one that is miles by miles in size. For example, a small asteroid, only fifty meters in diameter (164 feet) could be blown into a bubble two football fields in diameter with walls two feet thick. These walls are thick enough to protect from most meteorite impacts and space radiation. However, this is too small to allow us to develop a suitable artificial gravity by spinning around its own center. All of our residents would get too dizzy. But by using two of these "tiny" spheres, connected in a dumbbell arrangement, we can twirl it about the center between the ends of the dumbbell and achieve the desired gravity.

By putting floors in these spheres, we can have a large surface area for living space, for recreation, and for growing crops. We estimate that 500 acres of space could be available for these purposes in the ends of our dumbbell. Some of this would be sacrificed to make "open" spaces where several floors would be cut away to give a feeling of expansiveness, like the atrium in a shopping mall. This plan will make it practical to try out our hollow world scheme long before we have the capability to build as huge an artificial world as the one we were just describing.

## Life in a Space Colony

And, we will need to build a subscale world in order to develop the skills needed to live together in such a close community. Returning to the earlier discussion about how often to cause the rain to fall illustrates the point. This is the other side of the coin that many dreamers sometimes fail to see. We can have a physical paradise with the whole of our environment controlled exactly as we wish. But what do we wish and what do our neighbors wish and will these coincide? Power brings responsibility and the necessity for making decisions about everything within your sphere of power. If you have complete control over your physical environment, you also have complete responsibility.

The inhabitants of the hollow planetoid will have a physical paradise to the extent that they really know what they want as individuals and can agree as a group. And that brings us to the much more important question of the social and political life of the planetoid world and the advantages or disadvantages of such an existence as compared to our own.

The planetoid colony would necessarily be highly organized, with the cooperation, specialization, and interdependence of the inhabitants approaching that of the cells of the human body. The happiness and safety of the colonists would depend almost entirely on the degree to which they could achieve this social harmony. This is, of course, true on Earth also, but not to the same extent. The closed-cycle society would be far more sensitive to aberrant, antisocial, and destructive behavior.

Because of the security requirements for a highly disciplined society, some people have concluded that the space colony would necessarily have a totalitarian government—a *1984* "Big Brother" society complete with brainwashing, secret police, and the like. According to this theory, people would be forced into their tiny compartments of performing unrewarding, routine work with any deviations severely punished. However, this does not seem reasonable or probable. While some misguided governments might try to set up such a colony, it is doubtful that it could succeed for long. Like all the rigid totalitarian regimes of the past, it would contain the seeds of its own destruction. Since humans naturally resist compulsion, forced cooperation cannot be as productive as free cooperation. Also, the creative talents of the workers are stifled if people are forced to conform to the rigid plan of the government. But the most serious defect is the inherent instability of such a system. The best hope to prevent such a dire prognosis would be for a United Space Charter agreed to on Earth before any asteroid colony is settled.

Compulsory conformity cannot be effective unless erected on a basis of physical and psychological compulsion. But this implies the possibility of physical resistance, or violent rebellion. Violent conflict within a delicately balanced closed-cycle system would be catastrophic. While the necessity for specialization referred to earlier might suggest a workman doomed to a life of routine drudgery, such a situation would be unlikely. Routine work of all types would be handled by machines, and it would be primarily creative work which would be performed by the human inhabitants. Humans are basically independent and creative when not overwhelmed by group pressures to conform, and it would be their creative talent which would be needed in such professions as teaching, law, medicine, science, engineering, architecture, music, painting, writing, landscaping, military, statecraft, entertainment, etc. For the skilled but less creative there would be ample work in maintenance, repair, service, distribution of goods, etc.

There would be a much greater demand for teachers in such a society and also for members of the other professions. The greater material wealth of the society would be channeled into a higher percentage of doctors, lawyers, teachers, scientists, etc., reaching perhaps three to four times the percent in the United States of the 1990s. Thus, there would be one teacher for each ten students rather than for thirty or forty students and the level of education would be raised enormously. Likewise, the teachers would be freed from all routine tasks by machines and would be able to concentrate their time and energy on more important questions such as how to communicate new concepts and how to motivate students.

While it would be necessary to invoke a rigid discipline in times of emergency, full political freedom would be the rule. If people cannot express their opinions and air their differences in both verbal and written form, the bottled up resentments will eventually explode in violent rebellion.

The latest discoveries in semantics, game theory, the psychology of personality, motivation, and of optimized positive behavior would, of course, be available to the highly trained teachers of the new world. And the students would have been raised from birth in an environment of consistent, affectionate discipline with an optimum balance of freedom and direction. Thus, they would be emotionally equipped to respond vigorously, positively, and confidently to new learning experiences, and would learn rapidly and well.

The new world society, like any society, would have to be democratic to be stable. And stability would be essential for the safety of the whole society. And of course, a pure democracy with decisions reached through simple majority vote would not be adequate or stable. Representative government would be necessary as would constitutional protection of minority rights.

In short, it would appear that the carefully balanced closed-cycle society would *not* necessarily be totalitarian, and would, in fact, more probably have a government generally approximating those of the Western democracies. While it would be necessary to have a strong executive office and strict military regulations pertaining to some critical functions and situations, it would also be necessary to insure the individual political freedom and civil liberties. We have found in the West, and more recently in the East, that these things need not be incompatible.

In summary then, the question can be asked "What would be the purpose in setting up a colony in a hollow planetoid?" One way to answer this question is to ask another: "How would we describe the age-old dream of heaven or Utopia?" The dream worlds of heaven or Utopia which have fascinated people since the dawn of history have two basic characteristics, one involving the physical environment and the other involving human nature.

Our speculations regarding the perfect world or perfect society involve political and social theories, psychological and theological theories, and basic questions regarding the real nature of humans and their capacity for growth and improvement. The setting for our Utopia would not affect these things directly, although there could be important secondary effects. But physically, life on the captured planetoid could approximate closely our fondest dreams of paradise. Making it an emotional and spiritual Utopia is up to us.

## Note

1. D. M. Cole, "Extraterrestrial Colonies," *Journal of the Institute of Navigation* 7, nos. 2 and 3 (Summer/Autumn 1960).

# PART FOUR

# COSMIC COOPERATION IS POSSIBLE

# 17

# The Legacy of Columbus: 500 Years Later

Peter Pater, Astrogator, Lost his orbit calculator, Out among the asteroids . . . they rang the Lutine Bell at Lloyd's [of London].
—from *The Space Child's Mother Goose* (1959)

Five hundred years ago, a Genoese weaver's son came across a strange
map drawn by a Florentine geographer, Paul Toscanelli, changing the
course of human history on this planet. This map, based on the computations of the Egyptian astronomer Ptolemy and the Italian explorer Marco
Polo, had been superimposed onto a crude model globe by Toscanelli.
The Florentine geographer had used his globe to actually propose a westward voyage to the Orient as early as 1459 to Prince Henry the Naviga-
tor of Portugal. Toscanelli had been sent on an unsuccessful trip to the
bankers for a loan to subsidize the trip. He pleaded in vain that the westward run across the Atlantic to India would be shorter than an eastward
overland route.

Although his estimate of the distance from Portugal to China was off
by half the distance that we know it to be today (his five thousand miles
versus the 11,166 true airline miles), Toscanelli did spark an idea that was
still kicking around the Portuguese court when the Genoese, Christopher
Columbus, got to Lisbon through a fortunate shipwreck twenty years

later. He eventually laid eyes on Toscanelli's map when he fell heir to his Portuguese father-in-law's personal papers.

Armed with this document, Columbus proposed to King John II of Portugal, that he, Columbus, be allowed to test Toscanelli's theory. Failing to convince the monarch, Columbus next tried to peddle his ambitions to the British and French. Getting nowhere with them, he tried the Spaniards as a last resort.

Columbus did not have persuade Ferdinand and Isabella that the world was round, like a ball, since most of the schools of his time taught the globular nature of the planet. Instead, his real problem was to demonstrate to them that the ships then available were capable of reaching their hoped-for Oriental destination by sailing westward with enough food and water aboard so that the crews could be guaranteed a safe return home. Although the Portuguese had, in the latter half of the fifteenth century, developed large and sturdy vessels fit for the open ocean which had already circumnavigated the African continent before Columbus made his pitch to Ferdinand and Isabella, the Spaniards were not in such a fortunate position. Their ships were small and less seaworthy.

In a sense, our problem may be compared with that of Columbus. Our spaceships are primitive compared to what they will be a decade from now. But does that mean that we must wait until better ships are available before we venture into the New Ocean to seek our fortunes there? Furthermore, we have one distinct advantage over Columbus in our proposed voyages to the islands in space. We know they are there and we even know where many of them are located. The "Admiral of the Ocean Sea," on the other hand, miscalculated his target due to faulty maps and discovered the islands of the Caribbean instead of the American or Asian continents.

In one respect, though, we are still in a position similar to that of Columbus. Although our future space navigators will know where they will be going once they are launched, there will be many unknown hazards which may confront them en route to their space destination that may parallel those which faced the Italian explorer in 1492.

Once we have successfully sent out a series of unmanned space probes, first to the nearest asteroids that approach us, and then to the vicinity of the planetoid belt, we can start making concrete plans for the first manned flight to one or more of our selected minor planets. With the accumulation of valuable research data sent back from soft-landers which have descended safely to a preselected planetoid target, we will have answered many of the difficult questions about these small cosmic bodies before our adventuresome astronauts are launched on their first crewed missions beyond the moon.

At the end of the fifteenth century, when other European sea captains hugged familiar coastal shores, ready to scoot into any European or African port at the first sign of a storm at sea, Columbus and his three ships dared to probe the mysterious unknown waters of the Atlantic—a feat of incredible courage and vision. He stretched the horizons of Europeans and of humankind in general by opening the collective eyes and raising the level of consciousness of thinking men.

In the process, he helped to dispel the Ptolemaic, geocentric conception of the Earth forever and set the stage for the process that would lead man into space 435 years later. The New World colonization in the sixteenth and seventeenth centuries, triggered by the journeys of Columbus, led to the ascendancy of the West for the next five centuries, and the leadership of human exploration of the cosmos.

## Space Targets for Tomorrow

The great space controversy of the 1990s continues to rage around such questions as "Why go into space?" "Should we send people or instruments?" "How urgent is this new exploration?" "Where should we go after the moon?" We are particularly concerned with the last question because we believe that the next major target for human exploration should not be the moon, or Mars, but the planetoids. First, we need to learn more about their nature in the event we ever have to deal with a threat from one of them.

But, on the positive side, once we land on these minor planets, we can learn to exploit their hidden potential, including mining propellants, radiation shielding materials—and eventually, even platinum and other precious metals. We can even hope to capture them and return them to Earth orbit, where they can be utilized as natural space stations. A planetoid space station would ultimately be much less expensive than a large, artificial space station laboriously constructed in orbit piece by piece, since a planetoid can provide a plethora of materials which we would otherwise have to carry into space from Earth.

The early explorers of the New World had to carry with them just about everything they needed to cross the wide Atlantic Ocean. They could depend only on air and some water. (Our astronauts have to carry even those along.) Their wind-powered sailing vessels can be compared to our future interplanetary ships, coasting without thrust along most of their cosmic journeys.

Likewise, in the first days of crewed exploration of the solar system, it will be necessary to transport all the rocket propellants, air, water, food, and other supplies from the surface of the Earth at great expense. Consequently, use of crewed spaceflight for purposes other than exploration has been considered impractical or impossible by many scientists.

However, this phase will not last very long, or humankind will sentence itself to a starvation diet of resources in a solar system abundantly supplied with all manner of materials waiting to be exploited. We need not fear depriving another sentient life form of anything they need, because we are now pretty sure that we are the only intelligent life in our solar system. Already, engineers are making plans to utilize the resources of the moon and Mars to provide propellants for the return trips of early explorers.

As stepping stones to the planets and the stars, the planetoids can provide a valuable logistic support out to the far reaches of our solar system for setting up large interplanetary expeditions, bases, and then colonies. How well we succeed with this challenge of the planetoids will determine how soon we will be able to progress to the all-important and challenging phase of the exploration: exploitation and colonizing of the planets in our solar system. It will also help to chart America's future destiny in the cosmos.

At present, NASA's major interests for future crewed space projects lie primarily in three basic areas: an Earth-orbiting space station, a voyage to the vicinity of Mars, and a small, permanent base for lunar exploration. An early command decision on these programs is essential since the establishment of a lunar base or a crewed mission to Mars may take a lead time of a decade or so. Therefore, space events will be dictated largely by the choices made now, during the 1990s. At this writing, the Earth-orbiting space station, International Space Station Alpha (ISSA), has been selected by Washington as the major thrust of our space program in the first part of the twenty-first century.

In mid-1963, Dr. Joseph F. Shea, former deputy director of NASA's

Office of Manned Spaceflight, said, "The manned Martian program, I think, will be done primarily because of the exploration, not because it is economically feasible or because of an overriding necessity to go there. It will be done purely and simply because of the adventure and exploration. *It is the next challenging goal and focus for what we are trying to do.* [Emphasis added.]"[1]

We would now disagree with that assessment. It has been over twenty-five years since lunar exploration was stopped by politicians as being too expensive, and the public has not raised much of an outcry about this policy. We submit that the next phase of human expansion into the solar system must be expected to pay its own way over a reasonable time.

Columbus sailed to discover a shorter trade route to the Orient, and the economic potential of that was obvious to the sponsors of his venture. Cortez came to the new world for "Gold, God, and Glory." Our voyages to other parts of our solar system will have to be as economically self-serving, or they will not be sponsored either by government or by private ventures. There is already sufficient economic potential to go to the asteroids, and unlimited spiritual and cultural values may arise from such an enterprise.

Moreover, we need to remember that what Columbus *really* discovered was not a trade route, but something unexpected, a whole new continent with vast resources and native cultures. And the greatest benefit of his find, we believe, is not the material wealth that the new continent brought forth, but the new ideas and the culture of freedom that developed there. Those ideas of freedom have now spread to much of the world, and the end of that process is not in sight. Who knows what we may discover by venturing forth and colonizing new worlds?

It is important to draw some parallels between the New World Columbus discovered and the islands in space that we know are there waiting for us. In the beginning, it was expensive, difficult, time consuming, and dangerous to come to the New World. Further, once the settlers were here, they were very isolated from the Old World they had left behind. It will be so, at first, with the asteroid communities. Pioneers will have to fashion their own form of community to deal with the situations they find.

In America in the eighteenth century, there were vast lands unsettled to which people could move and make their own way, something very different from the situation in the Old World. The result of this frontier was the development of the American culture. Our openness, our willingness to take risk, our belief in the freedom and dignity of each individual came from this frontier spirit. We are in grave danger of losing our adventurous nature as Americans, because now there *is no frontier.* The asteroids offer us a similar frontier of almost unlimited opportunity, and there are no native humans to be exploited in the process!

When the English settlers came first to Roanoke Island, and then to Jamestown, they came to stay. They had no illusions about flying home to visit the relatives they had left behind. They could not imagine even a steamship, much less an airplane, that would make travel between the Old World and the New fast and convenient. In similar fashion, the pioneers who choose to settle the asteroids will not be able to imagine the new technologies that will someday link them to the planet of their origin.

## New Inventions from the Asteroids

The steam engine was invented in England. It was first used in industry, and then for motive power in England in primitive trains, but it was an American who first harnessed it to boats. The steamboat was born, serving humanity well for many years before other motive power, such as more efficient trains and then automobiles, supplanted it. We expect that the same pattern will be followed in the remote settlements of the asteroids. Of course, we expect that these asteroid colonies will be linked to "home" via radio and television, so that we will learn of new developments as soon as they occur. There is no way to guess how much humanity will benefit from these new developments.

We cannot pretend to predict what developments will occur in the planetoid civilization, but it is fun to speculate, based upon the needs that we can foresee, the opportunities presented by vast open spaces and abundant raw materials, and the science that we know today, which *may* be developed into usable technologies by our grandchildren living in the asteroid belt. One example may be the development of new propulsion systems. We mentioned the Hohmann transfer earlier as the most efficient orbit transfer currently known for moving from one orbit to another. Unless people develop incredible patience living in the asteroid belt, however, this technique will not be used much for human transportation because the travel times involved will be too long. The time needed to go from one asteroid to another, even one nearby, will be five years or more. So we expect the colonists to use less efficient orbit transfers to cut travel time to visit their neighbors, even though they will not "drive into town for a loaf of bread" very often. Thus, there will be a major need for space transportation that uses only local resources, making it affordable.

On some asteroids water will be available in quantity, bound in the very rocks that make up the asteroid, but on others, it will be quite scarce, so we would expect water to become a valuable trade commodity. It may very well be too precious to waste as rocket propellant. (Note, however, that the situation may be quite different on extinct comets, where water is so plentiful that it may well be consumed in vast quantities, primarily to move the remaining water to places where it is greatly needed.) This need for water, taken together with abundant supplies of aluminum and oxygen in the asteroids, may give rise to some novel rocket engines, in which the main propellants are aluminum and oxygen. Such a combination, under current technology, does not have the performance capability of our present hydrogen and oxygen engines (and note, both of these propellants can be made from water), but if water is scarce and these elements are abundant, the colonists will use what they have.

Another possibility is one that we have mentioned earlier: beamed power propulsion. In this concept, the rocket power source is separated from the rocket thruster, and a "working fluid" is used to provide the propulsive jet. In this kind of electrical engine, many kinds of materials may be used in the jets, since they do not depend upon the chemical or thermodynamic properties of the material. What will be important for this scheme to be effective, however, is the use of very narrow energy beams, so that most of the energy that is beamed from the transmitter to the rocket will arrive.

We already know how to make a narrow beam of microwaves. We know how to generate microwaves with reasonable efficiency, and we know how to convert them back to electricity at the receiver with very high efficiency. What is needed to make this scheme work is very large transmitting antennas. The space to make and use these is certainly available to our asteroid colonists. Consider, for example, an antenna a half-mile in diameter, to transmit energy at S-band frequencies. (This is the kind of microwaves used in a microwave oven, so we have a lot of experience in making and using them.) Such an energy beam could carry for long distances without diverging very much.* If every sizable asteroid colony had such a transmitter, it could launch and retrieve spaceships quite effectively. How this technology will be useful back "home," however, we cannot even speculate.

Such a technology would make travel among the asteroids far more practical than we can imagine today, and the colonists will have the motive, the resources, and the space with which to develop such a system. By the time there are several outposts there, however, they will devise much better ideas, making the suggestions we offer here seem primitive. But our purpose is not to describe how things may be done in the future, but to show that many advances over today's technology are possible, and we expect them to occur in due course.

We would note that as part of the development of this technology, large space structures, along with the means of building them, will need to be developed. Once this technology is in hand, new applications that we cannot imagine will no doubt appear.

## Feeding the Colonists

The one problem that was *not* faced by settlers to the New World was the problem of air, pure water, and food. They knew before they arrived that there was air, clean streams, and obviously, lots of growing plants. Our space colonists cannot depend upon any of these items. The asteroids are airless, the water is tied up in the rock, and they are so far from the sun that there may not be enough sunlight to grow crops even after the settlers make soil out of the raw materials on the asteroid. We will not have a thriving human civilization among the asteroids until we solve the engineering and biological problems involved in closing, almost completely, the food-air-water cycle that Mother Earth has so nicely arranged for us here.

---

*For the technically inclined, S-band is about 2.5 GHz. Therefore, a 1,000 meter antenna would have a half-power beam width of about 150 microradians. In other words, over 100,000 kilometers, the beam would only spread fifteen meters, so it is practical to intercept most of this energy with a receiver. This leads to practical solutions for spaceships plying the asteroid trade routes.

This problem is not unique to the asteroids. The same problems exist on the moon, on most of the solar system moons we know about, and—to a major extent—on Mars. Even for space voyages of several years, it would be most advantageous to have a closed life support system. This being the case, one would expect to find that this problem is being addressed by our national space agency, NASA. Sadly, you would be quite disappointed in your expectation. NASA has done very little to address this problem. It would appear that, because they are not addressing the problems that must be faced for this to happen, they do not expect humankind to leave Earth for space colonies for many centuries to come.

We believe that a careful and objective calculation of the payoff per dollar would place the planetoids at the top of the list of targets to be explored after the moon. They should certainly rank higher than any planet except perhaps Mars, and probably higher than the Red Planet itself. Meanwhile, if we do choose to go the planetoids, it would be best to go together rather than to go it alone, since the threat to Earth affects all citizens of our planet. As the late ambassador Adlai Stevenson put it, all the nations of the world are passengers on the Spaceship Earth.

**Note**

1. Joseph F. Shea, NASA Press Release (Washington, D.C., June 15, 1963).

# 18

# Capturing a Planetoid: Can It Be Done?

Someday, we will be able to bring an asteroid containing billions of dollars worth of critically needed metals close to Earth to provide a vast source of mineral wealth for our factories.
—Lyndon B. Johnson, in a speech delivered at the Seattle World's Fair, May 10, 1962

Historically, "capturing a planetoid" has involved only an astronomer's capturing it on paper or in a mathematical equation describing its orbit. Once the orbit characteristics were known with sufficient accuracy, it could not "escape" or become lost. Thus, it had been "captured." Of course, something quite different is in the context of the above quote from Lyndon Johnson—the actual, physical acquisition of the object. It is amazing, in retrospect, that a politician could imagine this thirty-five years ago.

## If God Had Intended ...

The reader might well have a strong negative emotional reaction to the suggestion that humans should modify the natural orbit of a minor planet.

It might be considered as the wildest kind of nonsense or even sacrilege. It is all very well to talk about flying to other worlds and even colonizing them, some may say, but changing their orbits? That would be desecration of God's universe!

Unfortunately for this kind of reasoning, its logical conclusion is that we never should have made fires, cut down trees, mined the Earth, or erected dams and bridges. In fact, we should have adapted to our environment—God's universe—rather than having the temerity to try to change it. But if we had adapted to our environment rather than trying to modify it, it is difficult to see how we would have differed from the dolphins, the gorillas, and other intelligent or semi-intelligent animals. If we did not try to understand, to change, and to control our environment, would we be humans?

In any case, modifying the solar system is no different in principle from modifying the Earth. The timing or scheduling of such attempts and the care and caution exercised are subjects for reasonable debates. But the eventual desirability and feasibility of exercising some control over the courses of the planetoids, icebergs, or rivers, is difficult to dispute.

Finally, we hope our readers agree that if an asteroid that could cause serious damage to our planet, perhaps to the extent of the annihilation of our species, came our way, we should do something about it! We would destroy it or divert it from such an impact. By this reasoning, we should have no more compunction about mining an asteroid, or moving it for our convenience, than we would have about mining iron or coal from the Earth. In fact, we will cause less harm to nature by mining the asteroids, since there is no wildlife there to be disturbed by our activities.*

## Is It Really Possible? . . . A Proposal

While it might seem at first glance that capturing a planetoid would be quite beyond our capabilities in the foreseeable future, note the following four points:

---

*NASA has spent vast sums of money to assure that spacecraft (especially those bound for Mars) carry no Earth life with them. These efforts indicate that long before we start moving asteroids around we will be quite sure they contain no life.

1. There are many planetoids smaller than the large objects that were first discovered. We have seen some as small as 100 meters, and there may well be smaller ones, although these may not be large enough to bother with. For one thing, we can expect asteroids that size to be stopped by the atmosphere unless they are mostly iron. Unless they are iron, their value as raw material may not be worth our effort to capture them. We should, however, be able to pick out any size object we wish for our first experiments.

2. We need not capture an entire planetoid, since it should be possible to split off the desired size chunk using nuclear explosives.

3. Nuclear explosives have such a great potential for certain types of space propulsion that it is really difficult to comprehend the magnitude of what can be accomplished before the actual numerical calculations are carried out. Some of these calculations have recently been done by the scientists and weapons engineers as part of the asteroid deflection workshop held at Los Alamos National Laboratory in 1993. We will call upon these computations shortly, since they are more credible than any we could offer.

4. The world has a vast surplus of nuclear weapons at present. Because of the ending of the Cold War, the superpowers have agreed to reduce their nuclear arsenal. We propose that these unneeded weapons be placed under international supervision to be held for the future when we are ready to remodel the solar system to more closely accommodate our needs. This will be cheaper than destroying such materials and will save us the trouble of building new ones sometime in the next millennium when we will need them to move asteroids, or for other feats of cosmic engineering of which we have not yet conceived.

## Asteroid Candidates for Capture

When Cox and Cole wrote *Islands in Space* in 1964, they were perhaps the first to seriously suggest that a planetoid be captured and placed into Earth orbit. Most "scientists" of the period scorned the idea and dismissed the authors as impractical. Cox and Cole were simply ahead of their time by many years. Now, even respected scientists at government laboratories discuss such ideas and publish papers on the desirability of doing this.[1]

In 1964, Cox and Cole wrote,

One of the most interesting groups of yet-to-be-discovered planetoids might be called the Earth orbit group—with orbits roughly similar to that of the Earth around the sun. They would be on roughly circular orbits with no more than five to thirty million miles difference between their perihelia and aphelia and, hopefully, would have only small inclinations to the Earth's orbital plane. Actually, it is not correct to say that there are no known members of this group, since several artificial members exist.[2]

These "artificial members" were several Russian and American space probes that either missed the moon, or were deliberately sent out to investigate the near-Earth space environment prior to the lunar landings. As an example of this, *Pioneer 4,* launched on March 3, 1959, missed the moon by 35,000 miles and went on to become an artificial asteroid. It is thought to have an aphelion of 105.8 million miles, a perihelion of 92.7 million miles and a period of 1.07 years. A return visit of *Pioneer 4* to the Earth was not made until fifteen years after its launch date. (While the Earth made fifteen trips around the sun, *Pioneer* made fourteen.)

This illustrates an important point about orbit commensurability. If a minor planet is detected as coming quite close to Earth, and its period is determined to be exactly 1.01 years, then a century and a year would pass before the object would again pass through the same point while Earth is nearby, although it would cross the Earth's orbit each year. Because close approaches may be so infrequent, it is especially important that we find as many such planetoids as possible, since a close approach on this orbit may mean an impact with Earth on the next. Of course, the COTE may be moving away from us, and will not be a threat again for a long time. But until we track it, we will not know in which direction it is heading.

No natural members of this "Earth orbit group" were known to exist in 1964. The first was discovered in 1992, and it is known only as 1992 JD. It very closely matches Cox and Cole's description, written thirty-five years ago, except that the inclination is 13.5 degrees, larger than desirable. This asteroid has a period of just over a year (1.01 years) and its perigee (the point at which it is nearest the object it is orbiting) can be closer to Earth than the distance to the moon. It is estimated to be about thirty yards across; about the size of a large house. It is not listed in the tables that show all of the close approaches in the next twenty-five years (Appendix D1), but that is just fine. We will not want to tackle capturing an asteroid for a while, and when we go after the first one, we will want one smaller than 1992 JD on which to practice. There are a number of other asteroids, most discovered only quite recently, that are also suitable candidates for capture.

There are also faint clouds which have been observed at one of the Earth Trojan points. The composition of these clouds is unknown, but they might well contain objects of the sort considered here. Very small velocity changes would be sufficient to bring an object from these clouds (or any of the Earth group) into orbits which will pass close enough to the Earth for capture.

Of the 160 Earth-crossing Apollo asteroids listed in Appendix D1, about a dozen could be "captured" using technology basically available today. The largest in this group is about three hundred yards wide. We may confidently expect more with these characteristics to be discovered in the near future.

## 1993 BX3: Come to Mother Earth

The largest asteroid in this "Earth orbit group" is 1993 BX3. It has a period of 1.65 years and has an estimated mass of 50 million metric tonnes. Its orbital parameters are such that it *could* hit the Earth in the near future. If it did, it would make the devastation at Tunguska look like child's play. We could try to move it such that impact will never occur, but it would be easier, and much more profitable, to capture and bring it into Earth's orbit for our future use.

To do this, we will give the asteroid a small velocity "kick" at apogee to bring it under our control. We will deliberately steer it close to Earth, close enough, in fact, that fine steering could bring it to an altitude in the upper atmosphere so that it would be captured into an Earth orbit by atmospheric drag.* (We will need to also give it another kick at its first

---

*The closest approach needed for this kind of a maneuver is around fifty to sixty kilometers above the surface of the Earth, far above where any airplane flies.

apogee above Earth so that on subsequent orbits it stays *above* the atmosphere and remains in a stable orbit.)

The exact size of the required velocity change will depend upon when we choose to do this maneuver. Since the period of 1993 BX3 is 1.65 years, it is constantly changing its position relative to Earth, and hence details of the capture procedure such as the exact velocity to be applied at apogee to steer it to our desired destination, also change. However, we can be assured that the velocity change will not be more than a few tens of meters per second. (Ten meters per second equals 22 miles per hour.)

But how do you move a mass of 50 million metric tonnes twenty-two miles per hour? One way is to use a series of nuclear bomb blasts on the surface. These will act as a crude rocket by ejecting thousands of tons of asteroid to give a jet propulsion effect. This will be very efficient, although we have such an excess of nuclear weapons of which to dispose that we need not be concerned in the slightest with engineering efficiency. The survival of a whole planet is at stake here!

This very proposition of how to move an asteroid was, as mentioned before, the subject of study by the experts in the field: the weapons scientists and engineers at Los Alamos. According to their report, to add ten meters per second to an asteroid of this size will take between 500 and 1,000 kilotons of nuclear explosive.[3] We will want to use several bombs to achieve this result, since the exact effect of each will not be known until it is tried. It is far better to explode one, measure the effect, and repeat as often as required to attain the desired results. If we use hundred-kiloton devices, we may expect each one to have a mass of about a hundred kilograms,[4] so we will not have an excessive load to transport to our destination asteroid.

Compare this procedure with using a rocket to add the needed velocity to 1993 BX3. This is not at all impossible, but it will require a lot more transportation capability from our spaceships. Imagine that we wanted to use the main rocket engines that power the space shuttle, the most efficient high thrust engines available to us. To add ten meters per second to our 50 million tonne asteroid would take the equivalent of all three space shuttle engines firing continuously for almost twenty-one hours. Engines to do that could probably be developed, but in that time, they would consume 112,000 tons of propellant. To haul that much propellant from Earth would be prohibitive, given today's technology, or even that expected of tomorrow.

The rocket approach would only make sense if we could find propellant raw materials readily available on the asteroid to be moved. In that case, it might be possible to take machinery to the asteroid, extract the raw materials (probably water) and convert them into rocket propellants. It might then be possible to use rockets to move an asteroid into an Earth capture trajectory. For now, however, nuclear bombs appear to be the better choice.

We need to be quite careful to assure that this massive asteroid does not come too close to Earth, or we will produce the very result we wish to avoid. All interplanetary missions provide for one or several "midcourse" corrections to adjust their "aim point" as they near their destination. By this technique, space engineers have been able to achieve remarkable precision in reaching distant planetary destinations. We will want to use such an approach here.

Suppose for example, we see that the asteroid is going to come one hundred meters too close. In the atmosphere this will make a large difference in the deceleration that the asteroid will experience. Too much and it might hit the Earth; too little, and it will not be captured. If we were to adjust the altitude by one hundred meters a month before arrival (about 2.5 million seconds) we would need to provide a steering Delta-V of forty micrometers per second. To add this much velocity to a 50 million ton asteroid will take five tons of propellant, if an engine like the space shuttle main engine were used. The engine would have to fire for about ten seconds to make the orbit adjustment. If we needed less correction, we could either shorten the firing time or wait until the asteroid is closer, so that the velocity would have less effect upon the trajectory. We can secure a miss accuracy as small as we desire, limited only by our ability to measure the orbit. Since we would put a transponder on the asteroid, the beacon it emits would allow us to accurately measure its range and velocity so that this procedure can be made perfectly safe for Earth. As a further fine control, we can use drag parachutes during the atmospheric braking phase. In this manner we can precisely control the amount of velocity we remove from the asteroid we are capturing. If the change is larger than we need, we can simply cut loose one or more of the several parachutes. This is another means of assuring a capture safe for Earthlings.

Once we have nudged the asteroid into its encounter with the upper atmosphere of the Earth, we will need to give it an "apogee kick," to raise the orbit above the atmosphere so that it will not gradually decay and fall to Earth. The Delta-V needed will be about seven meters per second (ten miles per hour) to put the asteroid into a seven-day orbit.

Once again, this could be done by large rockets. For an asteroid of this size, however, the rocket propellants required would be a daunting supply challenge if they were hauled up from Earth. If we had previously discovered an extinct comet and hauled many tons of propellants from it, rockets might be a viable proposition. Failing that, we would use another series of small atom bomb blasts to stabilize the orbit. Since these blasts would occur over 100,000 miles from Earth, we need not be concerned about their effect upon us.

We should note that 1993 BX3 has an inclination of only 2.79 degrees. If, as for many of the close-approach asteroids that we may wish to consider as candidates for capture, it were higher, we might wish to complete this inclination-changing maneuver in two or more stages. We have used planet fly-bys (which can be compared to a giant game of billiards which uses only gravity fields) for a number of years now to change interplanetary orbits. The current *Galileo* mission to Jupiter used three such encounters to get to its destination; one at Venus and two at Earth.

We can easily arrange such encounters with Earth to adjust asteroid orbits to make them easier to capture. The first such Earth fly-by would remove most of the inclination, and make an adjustment in the asteroid period such that it would return one or more times for further orbit improvements. Finally, when everything was to our satisfaction, we would schedule the entry maneuver to capture the planetoid into a permanent Earth orbit. With such a technique, asteroids of perhaps a few miles in diameter could safely be captured. Larger ones will have to wait until we have either more confidence in our prowess, or perhaps some new technology presently unimaginable.

Before we undertake a project as ambitious as capturing 1993 BX3, we will want to practice on much smaller asteroids. (We need to assert that the use of 1993 BX3—or the other planetoids mentioned here—is just an example for illustrative purposes. By the time we are ready to conduct an asteroid capture mission, we will have a whole new list of candidates from which to select the most suitable.) There are several such objects listed in Appendix D1 that have orbits available for such practice, all of which are small enough that if our skills are not up to snuff they will burn up in the atmosphere without causing much harm. One, 1992 YD3, will give us a chance to hone our skills in making asteroid plane changes, since it is now inclined 27 degrees. Of course, before we attempt to alter the orbits of any of these, we will need to conduct scouting expeditions to them first. They might be extinct comets which would crumble if we used nuclear explosives to attempt to move them, or they could be piles of rubble, which would do the same thing. If we find one of metal, however, it may well be worth our while to bring it "home," since it would have a mass of many thousands of tons of metal which we could use to build space stations, rocket ships, and other parts of our future space infrastructure.

We should note that our techniques need to be refined if we hope to capture extinct comets with their precious lode of water and other volatile materials. From what we know of comets, they are much too fragile to withstand the forces of atmospheric braking. If they fracture badly, this may allow the water and other ices we are trying to capture to boil uselessly away into space. However, we can solve this problem by enclosing such extinct comets in a giant plastic bag. They will still be crushed, but the bag will contain the gases so they can be recovered for our use.

## Other Interesting Asteroids

Thirty-five years ago Dan Cole described an asteroid called Hypotheticus that had what he regarded as an almost ideal set of orbit characteristics for a human visit. Following is his description of this ideal:

> If an object could be located with a perihelion closer to the Earth's orbit than any now known and with smaller inclinations to the plane of the Earth's orbit, the velocity requirements for rendezvous would be reduced. Thus we can hypothesize a planetoid with orbit inclination less than two degrees and a perihelion distance of less than one million miles from Earth. Also a period of almost exactly 1.5 years, or two years, would be most helpful.[5]

(The reason for the period requirement was so that it would reappear regularly.)

Well, this hypothetical body has never been found, but some come pretty close. For example, 1992 JB can come as close as specified, and has a period of 1.94 years. Unfortunately its inclination is 16 degrees, which makes it more difficult for humans to reach, but, as we said before, we can remove that inclination with one pass near, but outside, Earth's atmosphere.

Another most interesting object is the comet Wilson-Harrington. It was discovered in 1949 in the constellation Pegasus by Drs. Albert G. Wilson and Robert G. Harrington as they were beginning the National Geographic Society Palomar Sky Survey. L. E. Cunningham, who studied the photographs from the telescope, estimated the magnitude of the object as twelve and said, "All the images are strong and entirely asteroidal in appearance except for a small faint tail visible on the first two plates; there is no trace of coma."[6]

Initially, Comet Wilson-Harrington was estimated to have a diameter of 3.6 miles, a period of 2.3 years, and a perihelion of 1.028 AU. The corresponding figures for Hypotheticus were three miles, two years, and 1.01 AU. Unfortunately, the values for Wilson-Harrington were in error, and the comet was lost for a number of years. However, it has been recovered, and its orbit determined with enough precision that it warrants a number (instead of just a designation) in the asteroid charts. It is listed in Appendix D along with the other close approach asteroids.

Remember that it was initially seen to have a slight tail, strongly suggesting that it is an extinct comet. This means that there is probably a plentiful supply of water for space agriculture and for propellants waiting there for us!

## But, What Good Is It?

Perhaps the most important value of a captured planetoid would be its use as a stepping stone for longer flights into space. We will deal with this subject in the next chapter. However, there would be many other almost equally important applications of a second "natural" satellite of the Earth.

High on the list of importance would be the astronomical observatory and the science laboratory. Of course, such things are under development already and the many possible contributions from such undertakings have already been detailed elsewhere. But those developments and plans involve artificial satellites orbited, staffed, and maintained at enormous cost.

This relocated "natural" moon would have an abundance of raw materials for construction. We could provide walls thick enough to easily resist impacts from space debris, and shield against space radiation. With the abundance of building materials, we can construct large "farms" to grow our food and recycle our air. Remember, if it is a carbonaceous chrondite, we can expect to find large amounts (perhaps 10 million tons!) of water for drinking, for food crops, and for rocket propellants.

The laboratory or observatory on the captured planetoid would have the great advantage, therefore, of lower cost for construction, maintenance, resupply, etc., and could become a far more endurable or even attractive working and living site for scientists. In fact, the scientists might be quite willing or even eager to stay permanently (or at least for long periods of time) on the planetoid, thus reducing the cost of personnel rotation.

While cost saving would be the main advantage of the planetoid over the artificial satellite laboratory, perhaps that is the wrong way to consider it. The planetoid laboratory would mean that far more could be accomplished for the same investment, or that a given task could be accomplished in a much shorter time. Because of the ready source of supplies and raw materials and the much lower transportation costs, much larger laboratories could be constructed and staffed by many more scientists.

The possibility of mining the planetoids for valuable minerals has also been discussed, and here we will only consider some direct benefits which could become available to the average citizen. These include benefits to health, recreation, and the combination of pleasure and education characteristic of tourist travel.

## Cosmic Rest Homes

As a health sanitarium, the captured planetoid could offer all the attractions of the finest rest home in the world plus some additional benefits either difficult or impossible to attain on Earth. Such things as noise, dust, contamination in air or water, and unpleasant weather would be either eliminated entirely or controlled at a safe and pleasant level. Disease organisms would also be eliminated completely or controlled at a safe level. Colds, infections, and so on, would be absent or very rare. While these things can be controlled on Earth, it is very difficult to attain a satisfactory level of control. The pleasure cruise environment, to be described later, would provide a pleasant and interesting change for the patients and provide the best possible background for rapid recovery.

The unique aspect of the orbital hospital or rest home, however, would be very low gravity and consequent low weight of the people on board. On a planetoid, a person would weigh small fraction of an ounce. Thus, heart patients, arthritics, paralytics, and those crippled or enfeebled in any way from old age or disease could gain a new lease on life. Only the flick of a finger would be required to catapult a person from a chair and into a long, gentle trajectory through the air. Small wings possibly attached to clothing, or small electric propellers, would push people through the air to their destinations. The sick and feeble would be able to continue a mobile life while at the same time resting their bodies more completely than when lying in bed on the Earth.

Those who have been overimpressed by the difficulties, the stresses, and the dangers of our crewed space flights may question how an invalid could be safely transported to the planetoid hospital. They should not forget, however, that these tourist and health flights will not take place for twenty years or more. It is not anticipated that tourists or invalids will be subjected to the vertical, high acceleration take-off familiar to the television viewer. Rather, it is expected that aerospace planes will be developed during the next twenty years that will depart horizontally from a runway in much the manner of our present jet aircraft and fly into orbit. The passengers will be just as comfortable and safe as on a typical cross-country jet flight and could not even tell the difference without looking out the window.

## Space Tourism

This idea has been scorned by many "respectable" scientists and engineers ever since it was first proposed. NASA, however, recently negotiated a contract to explore the feasibility of the idea. Some future thinking engineers believe that this may be the use of space that will finally open the door to large scale operations in space. After all, tourism is the second leading industry on Earth. There is no reason to expect that people will not just as eagerly travel into space as soon as the cost of space travel permits (see chapter 12).

In the future it may cost less for a tourist to take a two-week cruise aboard a captured planetoid than it now costs to make a similar trip to Australia or the South Seas! The lucky vacationer of the future will stay at the orbit hotel where he will enjoy all the comforts and pleasures of the finest resort hotels on the Earth. But in addition, he will enjoy two types of pleasure not available to any of his less fortunate fellow vacationers who chose more conventional holiday trips on the surface.

First, the orbiting tourist will be able to enjoy a whole new dimension of sports, games, entertainment, and sensations made possible by the low gravity of the captured planetoid. Some of these will presumably be modifications and adaptations of the two-dimensional activities we enjoy on the Earth, but other yet-to-be-invented games will be particularly suited to the new three-dimensional freedom of motion.

Second, the tourist will have a completely unprecedented opportunity for sightseeing. While the average Earth-bound tourist on an ocean cruise has a monotonous view of sea and sky throughout most of his voyage, our orbital tourist will have a continuous and ever-changing view of the Earth below. The tourist on the several-week orbital cruise of the future, assuming a polar or near-polar orbit, will be able to contemplate, or study through binoculars, the entire varied and fascinating surface of the Earth and our moon.

Can humankind ever capture and convert an asteroid-planetoid for its own multi-uses? We already have the technology at hand. All we need is the imagination and the will to achieve this worthwhile goal of turning a deadly threat into a boon. The use of former weapons of terror to accomplish this feat would be a true "beating of swords into plowshares."

## Notes

1. J. G. Hills, "Capturing Asteroids into Bound Orbits around the Earth: Massive Early Returns on an Asteroid Terminal Defense System," in *Proceedings of the Near-Earth-Object Interception Workshop*, G. H. Canavan, J. C. Solen, and J. D. G. Rather, eds. (Washington, D.C.: GPO, 1993), pp. 243–50.

2. Donald Cox and Dan Cole, *Islands in Space: The Challenge of the Planetoids* (Radnor, Penn.: Chilton, 1964), p. 123.

3. John L. Remo and P. M. Sforza, "Near Earth Object (NEO) Orbit Management by Explosive Impulse Thrusters," in *Proceedings of the Near-Earth-Object Interception Workshop*, pp. 194-201.

4. G. H. Canavan, J. C. Solen, and J. D. Rather, eds., "Workshop Summary," in *Proceedings of the Near-Earth-Object Interception Workshop,* p. 204.
5. Cox and Cole, *Islands in Space.*
6. L. E. Cunningham, "The Discovery of the Comet Wilson-Harrington," *Journal of the American Astronomical Society* 220 (October 1994): 176.

# 19

# Stepping Stones to Mars and Beyond

The murky views which some scientists advocate as to the inevitable end of every living thing on Earth . . . should not now be regarded as axiomatic. The finer parts of mankind will, in all likelihood, never perish—they will migrate from sun to sun as they go out. And so there is no end to life, to intellect and the perfection of humanity. Its progress is everlasting.

—Konstantin E. Tsiolkovsky

A stream may be too wide to cross in a single leap no matter how we strain our physical resources. Or, it may be just so wide that we can make the jump with a great expenditure of energy and considerable risk. But if some flat stones happen to be available which we can place at short intervals across the stream, the crossing can be made with scarcely any effort or risk. The islands of the Atlantic and the Pacific served a similar purpose for the early explorers who could not carry sufficient fresh food and water for a safe crossing of the ocean. Could the islands in space be equally useful?

In order to sail the new sea of space we must have ships, and we must have propulsion systems to escape from the deep gravity wells that surround the planets and have thwarted men's dreams of weightless flight

256

since the days when men first learned to dream. Once in orbit, travel to distant cosmic lands is relatively simple and requires only a fraction of the energy needed for the original launch. Space sailing is not that different from ocean sailing in this respect, and, except for the need to sometimes trim our course to take advantage of the gravitational tides, we do not need to spend propulsion energy. All of the propulsion needs are at the beginning and end of our journey.

## Space Catapults

The moon may be such a stepping stone in space because there is a possibility of producing rocket propellants from its minerals and because catapults could be erected there to launch payloads into deep space. Arthur C. Clarke, the noted British writer, was one of the first people to discuss the value of the moon as a way station to deep space. He pointed out that its low gravity and lack of atmosphere make it an ideal point from which to catapult space payloads to more distant destinations. Likewise, the satellites of other planets and, especially, the planetoids could provide such energy shortcuts or stepping stones to the planets.

The guiding principle behind the space catapult is an electric motor, which is probably the most efficient known method of transforming stored energy into motion. If we could somehow achieve motor efficiency (about ninety percent) in space transportation, we would be close to the ultimate. Energy cost would then be such a small fraction of total costs that there would be little incentive to reduce them further.

An electric motor has another very interesting feature which makes it still more attractive as a space propulsion device—it can be run backward; that is, as a generator. If mechanical energy is supplied to rotate the armature, then electric power is generated. It may not be obvious, however, how an electric motor could be used to propel a spaceship. And, of course, the usual form of an electric motor, in which the motion of the armature is rotary, could not be used. What is needed is a linear motor, or electric catapult, and, in fact, such devices have been built and used for assisting the take-off of large aircraft.

In a linear motor, the armature moves along a straight track with the equivalent of a commutator (a device for changing the direction of electric current) reversing the polarity of the field coils as the armature passes by. Thus, the electromagnet of the armature is attracted by the field magnets in front of it and repelled by those behind.

The idea of an electric space catapult has been advanced considerably in recent years by a number of developments. The first of these was the development of the idea of space colonies by Professor Gerald K. O'Neil of Princeton University. He and Professor Henry Kolm of the Massachusetts Institute of Technology developed a magnetically driven catapult that can achieve high enough velocities to launch payloads from the moon. They called this concept a "mass driver," and O'Neil proposed to launch materials from the moon to build large structures, such as a large colony, in space. This concept is still being considered by the Space Studies Institute of Princeton, New Jersey. To avoid confusion with other magnetic launcher designs, this is sometimes called a "coil gun."

Another approach to a magnetic launcher is called a "rail gun." This approach is much simpler in design. A pair of straight, parallel rails, carries a very large electrical current which is passed from one rail to another behind the projectile (or payload) in a plasma current. The resultant force accelerates the payload along the rails until it exits at very high velocity.

Both of these concepts have some difficult engineering problems that stand in the way of their application to launching large spaceships at high velocity. The coil gun suffers from complexity; it requires many switches and a complex control system. The rail gun received considerable attention as a launcher for small projectiles during the Star Wars development effort, but it has problems with generating and switching multimillion ampere currents, and with erosion of the rails during repeated use. However, both of these concepts have demonstrated the principles needed for a high velocity space launcher. There is no apparent reason, except a difficult engineering challenge, why either one cannot be scaled up to serve as a space catapult.

Using a catapult to launch spacecraft from the surface of the Earth is quite difficult to do for two reasons. First, the Earth has a very high escape velocity, which makes a catapult large and expensive. Second, and even more challenging, the Earth has a dense atmosphere which will cause large drag losses, and very severe heating at the hypersonic speeds needed. Small, airless satellites and asteroids are ideal for this purpose, however. We will return to this idea shortly.

## Lunar-Based Rocket Propellants

The second reason to consider the moon as a filling station is the possibility of finding or making propellants there. To the best of our knowledge, there is almost no water on the moon; it is certainly not found in the regions explored by the *Apollo* astronauts. There is some evidence that there may be some frozen water in a deep crater at the south lunar pole. The floor of this crater is so deep that it never sees sunlight, so it is cold enough for ice to exist there for eons.

If water is found on the moon, it is almost certain to be only in limited quantities. Therefore, we have to consider how we want to use it. Should we convert it into propellants, or use it for agriculture to support lunar colonists? The inhabitants of a lunar city and the asteroid belt may have sharply divergent views on that! That contention could be enough to spark the first space war, unless we are wise enough to solve the issue before it arises.

We have mentioned the possibility of using other, quite abundant, lunar materials, such as aluminum and oxygen, for rocket propellants. This can certainly be done, although these are not as desirable as the hydrogen and oxygen that we could get from using water. In either case, the moon has a very significant gravity, with an escape velocity of 2.4 kilometers per second (5,300 mph) To use lunar propellants, we would have to lift them out of this lunar gravity well into space. While this could be done using a catapult, it may be easier in the long run to capture a planetoid into Earth orbit and use the resources it contains to assist our space travel rather than relying on the moon as a way station.

## Sailing the Extraterranean Sea

While launching ships onto the Extraterranean Sea does takes considerable energy, it is not nearly as difficult as many people suppose. The belief that "space is difficult" is fostered by people, such as the manufacturers of our present rockets, who are making large profits because space travel is now seen as expensive and problematic. In fact, it is *not* that much more difficult than similar tasks accomplished by more conventional Earth air transport systems.

This can be demonstrated by comparing the propellants needed for both intercontinental and orbital flight. A Boeing 777, the newest and most modern airliner available, uses about fifty pounds of jet fuel every hour for each passenger it carries. On a long flight, such as New York to Tokyo, this means that it will burn 750 pounds of fuel for each passenger. If an average passenger weighs 150 pounds, then the Boeing 777 uses five pounds of fuel for each pound of passenger to achieve its destination.

A single-stage-to-orbit rocket will require about ten pounds of rocket propellant for each pound that it places into orbit. Of course, a large part of that weight going into orbit will be in the spaceplane, and not just passengers, but this indicates the magnitude of the problem. In the future, after we learn to use the atmosphere as part of our propellant, as the "Orient Express" proposed by President Ronald Reagan planned to do, this fuel fraction will become much smaller. The principle difference is that the spaceplane uses all of that propellant in a spectacular few minutes, while the jet plane uses its ration of fuel over many sedate hours. From this comparison, we conclude that space flight is not all that difficult; it is just that we are at the "dug-out canoe" stage of space travel.

## Filling Stations in Space

Stepping stones or islands can greatly reduce the costs of transporting passengers and supplies through deep space and can someday even simplify the task of crossing the energy barrier from surface to orbit. We will consider some ways of transforming planetoids into stepping stones to the planets.

The most obvious way is to use the materials they contain as rocket propellants. We mentioned in chapter 9 that carbonaceous chondrites contain a significant fraction of water, so that we may expect to find water in many asteroids. We also know that there is a lot of silicon (sand) in an asteroid. Let us propose the following scenario. A group of entrepreneurs takes solar cell manufacturing machinery to an asteroid. The machinery is, in the mid-twenty-first century, almost fully automated. Thus, it can start with the silicon of the asteroid and make high quality solar cells in vast quantities in the vacuum of space. (It must be noted that making solar cells in a high vacuum and zero gravity environment is easier than making them on Earth.)

There is also a lot of iron in the asteroid, making it possible to build vast solar cell farms to catch the sunlight, feeble though it may be. Since these solar panels do not have to withstand being launched from Earth, are not subjected to winds or other storm forces (ice, rain), and exist in an almost weightless environment, they can be made huge in comparison with Earth structures. A mile by a mile on a side? No problem. Even as far from the sun as the main belt, such an array would capture 350,000 kilowatts of sunlight. If the solar cells are only 20 percent efficient (pretty shabby for the year 2050), these explorers will have 70,000 kilowatts of electricity available to them. If the want more, they can just enlarge their solar farm—there will be lots of silicon and iron left.

Now they can put their solar energy to use making rocket propellants from the water on their asteroid. They have enough power to produce more than a ton of liquid hydrogen every hour, along with eight times as much oxygen. Every week they can supply enough liquid propellants to fuel two space shuttles! That is more than enough to fuel great hordes of the deep-space craft that never need land on a large planet or moon. And tons of oxygen will be left over for sale to asteroid miners and prospectors. All this can be gleaned from the energy of the sun and the raw materials found in the asteroids.

So much for the vast expense and great difficulty of deep space travel. We say again, the obstacle in our space program is not engineering problems, but the lack of imagination in our space industry and its government leaders.

The question will arise as to where to put these filling stations. The locations are not obvious. An automobile uses fuel at a fairly constant rate; the farther you go, the more gasoline you need, so filling stations are strung out along the length of our roads and highways. In space travel, most of the trip is coasting, so the place to put filling stations is at the beginning and end of our popular trade routes. Since moving an asteroid is not easy, as we have seen, we will probably be willing to make reasonable detours to stop at filling stations. Where to put them will not be apparent until we begin to develop these "roads in space," just as the congestion that has developed in our own geosynchronous orbit was not apparent at the beginning of the space age.

An alternate approach to providing space filling stations is to use nuclear reactors to provide the energy for converting water (or other asteroid materials) into rocket propellants. On even small asteroids, the major problem of space reactors—the deadly radiation—could be readily overcome. The composition of the asteroid itself can be used to provide the nuclear shielding. The solution could be as simple a siting the reactor on one side of a small asteroid, and then just never going around the corner into the "line of sight" of the reactor. In this way, the radiation would go off harmlessly into space. (It is useful to note that a reactor is *not* dangerous until it has been first powered up. So it could be handled safely and installed by human crews, *until* it is first brought on line to furnish power.) Given the vast quantities of bomb materials now cluttering the Earth, this may be a means of putting this huge storehouse of plutonium to a practical peacetime use.

There is one further possibility that needs to be mentioned. We discussed in chapter 14 the use of beamed power as a means of driving our future spaceships. Using either solar or nuclear power, we could build power-beaming stations at strategic points in the asteroid belt to service the prospectors, miners, and settlers that may someday inhabit that part of our solar system.

## Earth Asteroid Filling Station No. 1

We described in the last chapter how an asteroid could be placed into orbit around Earth, using 1993 BX3 as an example. Now let us consider 1993 BX3's uses as a means of getting to Mars and to other destinations in the solar system. We did not specify in what orbit we would place this artificial moonlet. The easiest one to attain and, we believe the most useful, is a highly eccentric Earth orbit (HEEO). Such an orbit would come close to the Earth (say, 300 miles up) at its closest point, but travel out toward the moon at its farthest retreat from Earth.

At the point of closest approach (perigee), the velocity of the asteroid is just marginally less than escape velocity from the Earth. Specifically, for a 300-mile perigee and a seven-day period, the velocity at perigee is 23,800 mph. This is only 56 mph less than needed for a trip to the moon, and only 1,800 mph less than the velocity needed to go to Mars or to its offshore islands, the Martian moons. These velocities could be readily attained by rockets leaving our captured asteroid.

The other side of the coin is that it is more difficult to reach this velocity from an Earth surface launch. The circular orbital velocity at 300 miles is only 17,000 mph, some 6,800 mph less than our asteroid. With present technology, it will be very difficult to increase our Earth launcher performance that much. In fact, some skeptics claim that it will be many years before we can even attain the 17,000 mph with an single-stage-to-orbit launcher, much less almost 50 percent more.

This is where our filling station concept comes into play. Remember the entrepreneurs on an asteroid in the main belt, selling rocket propellants? You can be sure that they would get to our Earth-orbiting asteroid long before they went to the main belt. For one thing, the sun is ten times as intense at Earth, so their solar panels can be ten times smaller, or, they can produce ten times the amount of propellant from their space factory.

So, here is one plausible way that this scheme can work. Let there be a spaceplane to shuttle from our captured asteroid to low Earth orbit (LEO) and back with people and cargo. Actually, there should probably be two different vehicles, a space taxi for people, and a space truck for cargo. We will describe the operation of the space taxi, but the space truck will work in about the same fashion, except that it will be bigger, and could operate without the expense (and risk) of having a crew.

Using fuel manufactured on the asteroid, the space taxi departs from the asteroid docking port (hanger) sometime after apogee. Leaving at apogee would be the most efficient and economical way to operate, as the Delta-V that would be needed would be only fifteen miles per hour. However, it would have to provide accommodations for the passengers for a three-day trip. Hence, we will defer its departure until a day or less before the asteroid's arrival at Earth. This will increase the necessary Delta-V a little bit, but is much less of a penalty than providing meals and sleeping accommodations for several days. (The space truck does not have this limitation, and can use the more efficient trajectory.)

The space taxi will enter the Earth's atmosphere to slow down enough to rendezvous with the space hotel in LEO. It will pull into the "lobby" and discharge its passengers. The taxi is carrying all of the propellants it needs for the return trip. These were manufactured on the captive asteroid, which was one of the reasons to capture it in the first place. After the taxi picks up the outgoing passengers, it departs from space hotel to return to the asteroid. The passengers may be going there for a week's vacation, or they may be changing flights for the moon, Mars, or the main belt.

After it leaves the hotel, the taxi fires its rocket motors again to climb back into the asteroid's orbit. In a day or so it will fly into the hanger on the asteroid, and the passengers can be on their way. Note that *no* propellant from Earth was used for this trip, so the trip is no more difficult for the Earth-based transportation system than any flight to low orbit.

In order for the space taxi to perform all of these maneuvers, it will need to be about half fuel by mass when it leaves the asteroid. This compares favorably with the Boeing 777 passenger jet we described earlier which is about 40 percent fuel (maximum) when it takes off. Jet airliners weigh between 500 and 1,000 pounds per passenger, depending on size, range, and other factors. The space taxi will not need big wings or landing gear, but it will need an effective heat protection system to shield it from the heat of its flight through the atmosphere at Mach 25 to 30. We will call that a fair trade, and say that the space taxi will weigh 1,000 pounds for each passenger.

If this is so, then the fuel required for each passenger will be 1,000 pounds. If only ten percent of our asteroid is water, it will be able to provide the fuel for transporting 1,000 passengers per week for over *two hundred* years. And this is just one small asteroid, one that is mainly rock!

Of course, we will not want to use propellant only for flights to and from Earth. Rather, we will want to use some of our resources to travel on to the moon and Mars. Suppose that we use rockets to go on to Mars. It will take a velocity increment (Delta-V) of 1,800 miles per hour to do this, considerably less than required for the trip to LEO. Since this is so, the fuel required for each ton of payload boosted toward Mars is only 325 pounds. Suppose that we boost a 500-ton payload toward Mars at every opportunity (about every two years). Each mission will take 82 tons of propellant from our asteroid. In two hundred years, we will use only one-sixth of one percent of our asteroid's water for "side trips" to Mars!

But what can we do with a five-hundred-ton spaceship? Remember that the *Apollo* ships that went to the moon three decades ago weighed only fifteen tons, not including the propellant needed to go into lunar orbit and return. If we have the propellant waiting for us in a filling station at the other end, we do not need to carry that along. Of course the trip will be much longer, so we will need to carry more supplies, and we will probably want a larger crew.

The supplies now needed for an astronaut are about fourteen pounds a day for food, water, and oxygen, without recycling anything. Surely, by the time we go to Mars we can do at least some recycling, and reduce this requirement to eight pounds per day for each crew member. Now, for a 250-day flight, each crew member will consume a ton of supplies. So let us imagine a crew of fifty. They and their supplies will now weigh 55 tons. This leaves us 445 tons to build a spaceship; or thirty times what *Apollo* used for astronaut quarters and support. This should allow us to mount a very sturdy expedition to the Red Planet.

## Catapulting to Mars

But this rocket approach is not yet the best way to use the asteroid to enhance outerspace exploration. We will return to the idea of an electric catapult. It can readily provide 1,800 mph to our outward-bound space-craft. Both rail guns and coil guns have demonstrated velocity capability greater than 1,800 mph, although not for loads as large as we wish. If we build a catapult six miles long (in the weightlessness of space, using iron from our asteroid) we can launch to Mars without using any propellant at all! While such a structure might seem fantastically large if thought of as a single building, it would be quite small compared to the thousands of miles of railroads, highways, power lines, etc., which we have con-structed in the United States. And it need not be a solid structure, but rather a truss frame holding a network of cables and rails.

This asteroid accelerator would be used to send payloads into trans-fer trajectories to the moon, Mars, or or the main belt, where they could be caught by similar accelerators on similar orbiting satellites. Mars, of course, already has such satellites in orbit—its moons Phobos and Deimos—which are just the right size for this purpose. The ride on this catapult will be just a bit harder than a good carnival ride (3.2 gee for twenty-seven seconds), but nothing that most people could not withstand without much discomfort. (For people going to the moon for a low gee recuperation or living, because they are injured or frail, we can use rock-ets as a sort of "space ambulance.")

Of course we do not get something for nothing, even in this case. The momentum we have imparted to the Mars ship will have to come from our asteroid. In other words, the force needed to lift the ship up into space is created by pushing down on the asteroid.* This is not of immediate concern, however. The thrust from takeoff will slow the 50-million-ton asteroid in its orbit (at perigee) by 0.016 mph. If we don't correct this, it will eventually cause our asteroid to drift into an orbit that we do not like. But there are two easy ways to correct this loss without using rocket propellants, if we wish.

One way is to use the electric motor principle again. This time we extend a long tether from our asteroid. We pass a current (derived from either solar power or a nuclear reactor) along this wire. It interacts with the magnetic field of the Earth and provides a motor force restoring the asteroid to its original velocity.

The other way is even simpler: use the same magnetic catapult to launch payloads from the asteroid down to an easy re-entry velocity for return to Earth. The jet reaction of these launches will restore the momentum of the asteroid lost in launching payloads away from Earth. These payloads could be derived from resources on the asteroid, or they could be materials returning from the moon, or the asteroid civilization that we postulate for the middle of the next century. They could be useless rock, for all that it matters to the orbital stability of the asteroid, but we expect that there will be lots of freight returning from deep space by the time we have our new satellite in position.

It is plain that placing an asteroid in Earth orbit will open the solar system for our exploration.

## The Tale of Two Moons

Two of the most interesting of the islands of the new Extraterranean Sea lie close to the shores of the small "continent," Mars. They are so close that they must be considered to be well within its "territorial" limits.

The two small satellites of Mars, Phobos and Deimos (Fear and Terror), were discovered by Asaph Hall on August 11 and 17, 1877. This important milestone in the history of astronomy came only after many

---

*This seeming paradox is explained by Newton's third law of motion, which states that for every action there is an equal, but opposite, reaction.

years of painstaking search by Hall and other astronomers. According to Willy Ley, much of the credit should go to Hall's wife, since Hall had become discouraged in the search and was ready to give up. She urged him to continue searching even into the very unlikely region near the planet, and here he finally made his great discovery.[1]

These new satellites were very interesting to astronomers for several reasons. For one thing, they were small, much smaller than any other satellites known at that time. Phobos has been estimated to be about ten miles in diameter and Deimos about half that. Even more surprising however, was their close proximity to Mars; that is, the small size of their orbits. Deimos, the smaller, outer satellite, has an orbit radius of 14,600 miles and an orbit period of thirty hours and eighteen minutes. Phobos, with an orbit radius of 5,800 miles, is only about 3,600 miles above the surface of the planet. Its orbit period is seven hours and thirty-nine minutes. Thus, it revolves three times around the planet in a single Martian day of twenty-four hours and thirty-seven minutes. Since no other satellite was known to rotate in less than a "day" of its primary, the discovery of Phobos was both unexpected and challenging to the theoreticians. The Martian moons were interesting for another reason also. Their discovery had been predicted many years before the event (or perhaps we should say "prophesied," since the prediscovery "guesses" were not based on scientific deductions).

The great astronomer, Johannes Kepler, "guessed" that Mars had two moons as early as 1610, and Anton Schyrl, Kepler's contemporary, claimed to have seen them in 1643. However, the telescopes of that period were quite inadequate for detecting such small and distant objects.[2]

The English satirist Jonathan Swift referred to two moons of Mars in the section of *Gulliver's Travels* called "Voyage to Laputa." He described the orbit of the innermost satellite as being exactly three planet diameters in radius with a period of ten hours, and the outer satellite being at a distance of five diameters with a period of 21.5 hours.

Dr. Charles Olivier, an American astronomer, notes that the moons of Mars can never be seen through small telescopes and only under good visibility conditions through the largest. Also, he notes that Swift wrote about a century and a half before the discovery of the satellites and about a century before any instrument was made with which they could possible be seen. Olivier goes on to say,

When it is noted how very close Swift came to the truth not only in merely predicting two small moons, but also the salient features of their orbits, there seems little doubt that this is the most astonishing "prophecy" of the past thousand years as to whose full authenticity there is no shadow of a doubt. The nearness of Phobos to the planet's surface means that it can never be seen from areas in the polar regions. Its period of less than eight hours obliges it to rise in the west and set in the east.[*] In this latter respect, it is unique among all bodies in the universe, so far discovered. Yet Swift had this fact also included.[3]

If the remarkable "prophecy" by Jonathan Swift is unsettling to those who long for a well-ordered logical universe, a theory proposed by Soviet physicist, I. S. Shklovsky, is even more distressing. He suggested, as reported in *Astronautics* (December 1959), that the moons of Mars are artificial![4]

Shklovsky notes several unique characteristics of the Martian satellites in support of his theory: their very small size, their extreme proximity to their primary, the period of revolution of Phobos—less than a Martian day—and the equatorial plane of their nearly circular orbits. Shklovsky does not explain why he thinks the equatorial plane of the orbits is significant, except to claim that it rules out natural capture of planetoids as a possible origin.

If these "uniqueness" arguments were all that Shklovsky had to offer, he would probably have remained silent. However, he had noted something much more important. He believed that comparison of observations of Phobos in 1945 with others made early in the century showed that Phobos was speeding up in its orbit. It seems to be moving in closer to Mars as it would if atmospheric resistance were changing its orbit. This might not be important except that Shklovsky's calculations show that the observed change is far too great for any natural object. Only an object of very low density, like a balloon or a large spaceship, could possibly be affected so greatly by the tenuous atmosphere at that altitude above Mars. Thus Shklovsky concluded that Phobos must be hollow and must, therefore, be artificial.

---

*Because Phobos has a period which is shorter than the amount of time it takes for Mars to rotate upon its axis once, the moon appears to reverse the "natural" Earth order and rise in the west and set in the east.

The same issue of *Astronautics* which carried Shklovsky's theory also included a critique by Dr. Clyde W. Tombaugh.[5] Dr. Tombaugh, who discovered the planet Pluto in 1930, is one of the world's most careful and accurate planetary observers. He made hundreds of observations of Mars over a period of about thirty years and his first-hand knowledge of the planet can hardly be surpassed. He suggested that the observed orbit change of Phobos could be accounted for by assuming repeated impacts of tiny planetoids or meteoroids caught in the gravitational field of Mars, and scoffed at the notion that they were artificial. Nevertheless, the claims by Shklovsky caused a minor sensation at the time.

We should note, in passing, that the photographs of Mars obtained in recent years by our planetary probes show surface erosion that makes many scientists believe that Mars once had much larger quantities of liquid water than it does today; oceans may even have existed. If this is so, then the atmosphere of ancient Mars was probably also quite different, and capable of supporting life, as has been hypothesized, based on studies of a Martian meteorite published in August 1996. What happened to the water and the atmosphere has not yet been fully explained.

There is another objection to the theory of the artificial satellites of Mars which Tombaugh did not mention. If the Martians were so capable that they could build satellites ten miles in diameter, what could have brought about their demise? If the Martians could build Phobos and Deimos, they should also have been able to plant self-sustaining colonies on the planetoids, the satellites of the major planets, Mercury, and our own moon. In fact, it should have been easier to colonize the planetoids than to build giant orbital stations. But once their colonies were spread across the solar system, what could possibly have wiped them out? Why are they not still the masters of the solar system?

The photos of Phobos and Deimos, taken long since this theory was proposed, show them to be irregularly shaped objects, with craters and ridges just like the asteroids of which we have just so recently obtained pictures. Astronomers now believe the Martian moons to be captured asteroids, although they have not been able to explain how natural processes could have caused their capture. This seems to further discredit the notion they are artificial, but the pictures we have are not detailed enough to rule out airlocks to the interior of a hollowed-out asteroid.

Were these tiny islands used as bases by the Martians or by interstellar colonists? Will we be able to use them as stepping stones to Mars and the outer planets?

## Landing on the Off-Shore Islands of Mars

Dr. Ernst Steinhoff and others have argued that the best target for our astronauts, after the moon, is not Mars, or even the asteroids, but the satellites of Mars.[6] As noted, these objects appear to be asteroids that somehow ended up in orbits around Mars, so most of what we have said so far about asteroids should also apply to these moons. Since it should be easier to make a round trip to Phobos or Deimos than to Mars itself (because of the energy requirement for actual descent to and launch from the planet) such a trip could precede the Mars landing and be accomplished sooner, with less advanced vehicles and at a much lower cost. Steinhoff also argues that propellants extracted from the minerals of Phobos could be used to refuel vehicles which make the later and more difficult flights to Mars.

We disagree with that priority, preferring to explore the asteroids first, but we endorse the notion that Phobos or Deimos should be targeted for a human landing before Mars itself. As we have noted, we can expect to find the raw materials for propellant manufacture there. This would permit us to explore Mars in a far more robust way that we could if we had to carry everything with us from Earth.

It is probable that Phobos or Deimos can be more useful as a stepping stone to Mars that any known planetoid. Other planetoids may be more useful for spaceflight in general—not only for Mars, but for trips to the outer planets. Someday, we may eventually place a third asteroid satellite around Mars or move one of the present moons closer to the planet, to make it a better transfer point for Mars trips.

Farther into the future, we will want to construct electric catapults on one of the moons of Mars. The choice of which will depend upon their composition, as well as their closeness to Mars. When these catapults are in place, a spaceship from Earth will approach the Martian moon, be captured in the catapult, and reverse the process we described for leaving Earth. Then a round trip to Mars will be no more daunting than a trip to the New World was in the days of the Spanish explorers. It will take a long time, but the route will be known, and there will be no hostile warriors at the end of the journey.

## Mars Direct

We want to note, and applaud, the Mars Direct proposal advanced by Robert Zubrin of the Lockheed-Martin company.[7] His scheme involved building a large new space booster derived from hardware now used in the space shuttle program. One of these would be launched, without a crew, directly to Mars, carrying a nuclear power supply, a load of hydrogen, and a return rocket for a future astronaut crew. Once on the surface of Mars, the machinery would set about using the power from the nuclear reactor, the hydrogen, and the carbon dioxide known to be in the Martian atmosphere, to make propellants for the return rocket the mission had brought with it.

After telemetry (the transmission of information from the automated instruments) had confirmed that the return rocket was fueled and ready, a crew of astronauts would be launched to land on Mars at the site of their return spacecraft. They could explore for weeks and then return home. By flying two such large rockets at every opportunity, Zubrin proposes a vigorous Mars exploration program, leading in a few years to the first permanent Martian outpost.

This approach is vastly to be preferred to the plan that NASA prepared for President George Bush when he proposed, on the twenty-fifth anniversary of the *Apollo* moon landing, that we go explore Mars. The scheme proposed was nothing more than a plan to vastly enlarge the size of the faceless bureaucracy that NASA has become. It did *not* use any of the imaginative ideas that have been proposed over the years to make space travel easy and affordable. Instead, it perpetuated the image of an elite astronaut corps sustained by hordes of attending government workers. Whatever happened to the image of the intrepid explorer, out to challenge Mother Nature on her home turf?

## Notes

1. Willy Ley, *Watchers of the Skies* (New York: Viking, 1963).
2. Ibid.
3. Charles P. Olivier, "Mars," *Encyclopedia Americana*, Vol. 18 (Danbury, Conn.: Grolier, 1960), p. 320.

4. Cited in Don Cox and Dandridge Cole, *Islands in Space: The Challenge of the Planetoids* (Radnor, Penn.: Chilton, 1964).

5. Ibid., pp. 136–39.

6. Ibid., p. 139.

7. Robert Zubrin, *A Case for Mars* (New York: Free Press, 1996).

# 20

# Is Anyone Else Out There?

It is difficult to say what is impossible. The dream of yesterday is the
hope of today and the reality of tomorrow.
                                                    —Robert H. Goddard

The question if life exists elsewhere in the universe has bothered both scientists and philosophers for a long time, and we are just beginning to get
some concrete answers. The first question which must be asked is, "Are
there planets revolving around other suns?" We now have a definite
answer to this question, and it is *yes*! A second question has now also
been asked: "Does life arise naturally in the evolution of a planet if conditions are proper?" The answer to this is a resounding *maybe*.

Just recently, astronomers have discovered other stars that have planets
around them. They have not "seen" them yet, but have deduced them from
other observations. The technique is simplicity itself, in theory. When a satellite revolves around its primary (moon around Earth, or Earth around sun,
etc.) the satellite causes a small "wobble" in the path of its primary. The size
of this perturbation depends upon the mass of the satellite and its distance
from the primary. But, with very careful observation, this wobble can be
detected, and thus the presence of a satellite inferred. This requires very careful observation work by the astronomer, and is quite challenging in practice.

273

This answer to the question "Does life arise easily?" has come to us quite recently, when in August 1996, NASA and President Bill Clinton announced that they had strong evidence that there had long ago been life on Mars, in our own solar system, when the planet was warm and wet. Of course, some scientists are questioning if the forms discovered in the Martian meteorite (designated AH84001) were the result of life, or some other natural, but nonliving, chemical process. In either case, no one has seriously suggested the existence of any intelligent life in our home solar system, now or in the past. For that, we must still, we believe, look to other stars.

We need to note that NASA has always been quite careful about transporting life forms from one place in the solar system to another. For example, when the *Apollo* astronauts (*and* their rock samples) returned from the moon, they were kept in quarantine for some period of time. One reason was so that any moon microbe would not affect Earth. Even after it was certain that contamination was not a problem, there was still a very slim chance that some unknown, lifelike, microorganism might be found in these samples, and NASA wanted to be sure that such a microbe could not have been introduced on Earth after the samples were returned.

In similar fashion, extreme precautions have been taken to be sure that microbes from Earth were not introduced to other planets by our spacecraft, since one question that the space program is intended to answer is "Does life exist anywhere else?" This was particularly difficult for the 1976 *Viking* lander since it had to perform several experiments to test for the presence of life on Mars. It was a major challenge to keep the life-testing instruments free of Earth life. It has also been a challenge to build a spacecraft that could be completely sterilized before it left Earth, since the preferred method of sterilization was through heating. Finding spacecraft parts that could still function reliably after such treatment was both difficult and expensive.

With the new evidence that there may once have been life on Mars, we feel sure that NASA and other global space agencies will continue their efforts to avoid introducing Earth life onto another planet.

## Distant Planets

It is not surprising that the first planet orbiting another sun that was detected (in 1995) was very large and very close to its sun, 51 Pegasi. In fact this planet is so close to its sun that water would boil on its surface. Thus, scientists hold no hope for any recognizable form of life on that planet. In January 1996, the discovery of two new planets that seem more promising was announced. These orbit stars 47 Ursa Majoris and 70 Virginis. Both of these *could* theoretically be capable of sustaining life from the little bit that we know of them. All three stars with the new planets are within forty light years of the Earth and each is about the size of our sun.

The planet orbiting 70 Virginis is believed to be five times the mass of Jupiter, and is closer to its sun than Venus is to our sun. This planet has a "year" of 117 days. The planet orbiting 47 Ursa Majoris is thought to be smaller, merely 3.5 times the mass of our largest planet (Jupiter), and it is much farther from its sun; farther than Mars is from our sun. Its "year" equals about three of ours. Both of these planets are predicted to have surface temperatures that permit liquid water, which is considered essential for "life as we know it."

Since all we know of these planets is how much (and how fast) they cause a "wobble" in the motion of their sun, we are unable to speculate about other planets in their solar system, or anything else about them.

## A New Extrasolar Planet Search

In 1995 NASA announced a new initiative to find still more planets orbiting nearby stars. The plan, known as Exploration of Neighboring Planetary Systems (ExNPS), was described in a speech to the American Astronomical Society by NASA administrator Dan Goldin. The technique NASA proposes to use is quite different from the one that inferred the existence of the three previously discovered planets. They propose to build a monster *virtual* telescope in our solar system, out beyond Jupiter, where it is very dark. This telescope would be so big that they will be able to actually see ("image" is really a better description) planets around these stars.

Before we explain that, let us describe what is meant by a "virtual" telescope. A standard telescope forms an image at its focal point by bringing together the light photons of various colors in the proper relationship (in phase, to an engineer). For an ordinary optical telescope, this is done by the human eye in "real time" in "analog form." However, it is not essential to do it in this manner. If the phase information (color, if in the visible spectrum) is preserved, it may be converted into a digital format and stored indefinitely. Then later, in a computer, this phase information can be combined with other parts of the telescope to recreate the image.

Note that this process does not require that the telescope be all in one piece or even in one place. Provided the time at which the phase information is recorded accurately enough, the data may be taken from different, widely separated locations. Technically, this is called interferometry and has been used for a number of years now with radio telescopes. A very large array of separate radio telescopes has been observing the sky now for years using this technique.

This is relatively easy to do for a radio telescope, because radio waves are very long compared to light waves; at 300 megaHertz, a radio wave is one meter long. To use the "virtual" technique, it is necessary to record (and preserve) the phase information of a small fraction of one wavelength. Hence, to use this technique to form an image at radio wave lengths, the time of arrival of each part of the wave must be known to within, say, 0.1 of a nanosecond. (A nanosecond is one billionth of one second.) This will record the wave shape to within a one-thirtieth of a wavelength. Even higher accuracy would be nice, but we can work with these numbers. Atomic clocks can routinely provide subnanosecond accuracy, so the use of virtual radio telescopes is a well established practice.

However, for visible light, the waves are only half of one-*millionth* of a meter long, requiring that the timing accuracy be a million times better in order to build a virtual optical telescope. This has not yet been achieved, but appears to be possible. It is this principle (with the variation that short infrared wavelengths will probably be used, making the problem slightly easier) that NASA intends to use to build a large virtual telescope in deep space.

The details of how this will be done are still very much in the planning stage. However, it is very likely that the technologies being developed for the New Millennium program will also be applied to this program. NASA is thinking of launching three spacecraft together sometime between 2005 and 2010 to begin this exciting new exploration. Two of these spacecraft, separated by many thousands of kilometers, would probably make the observations with infrared telescopes. The third would combine the data from these to form an image and radio it back to Earth. If these are microspacecraft derived from New Millennium, they could be launched on a single rocket, and then use their ion engines to fly into the deep, dark space beyond Jupiter to make their measurements.

## Help from the Asteroids

This same technique can be employed and upgraded by using the asteroids. The "power" of a telescope, real or virtual, is determined by the amount of light (energy, radio waves) that it can collect, and by the distance from one edge to the other. For a "real" telescope, both of these are determined by the size or aperture. In a virtual telescope these two factors are quite independent. The virtual telescope that NASA is planning will have an edge to edge distance of thousands of kilometers, so it may be able to form images of nearby stars. However, the amount of light that it can collect will be quite limited by the size of the individual spacecraft telescopes. This may limit its ability to "see" small planets like Earth.

Once we have settled into the asteroids, we can build a much more powerful planet-seeking virtual telescope. First, we can use as a baseline, for instance, the distance from the Earth to our chosen asteroid. Instead of a separation of thousands of kilometers, we can have hundreds of millions of kilometers of separation. This would permit us either to see much smaller objects at nearby stars, or see large planets much farther away.

This increased separation will not do us much good, however, if we are limited to telescopes at each site that are too small to gather enough energy to form an image. Thus we will need much larger optics at the asteroid, and at Earth. (One probability is that the optics may be in orbit around the Earth.) The telescopes that NASA is discussing for their planet discovery program have an aperture of 1.5 meters. We will want to do much better, and increased size is feasible, for our asteroid telescope.

A factor of a million times in optical collecting area might be a good match for our many millionfold increase in baseline. This suggests a telescope with an aperture about a mile in diameter. Such a thing would be somewhere between difficult to impossible on a planet, because gravity would make it sag too much to be a good telescope. But in space, without a gravity field to worry about, use of such a telescope is quite possible. And it need not be very massive, so a very small fraction of an asteroid would provide all of the material needed to build it. With such an advance in astronomy, we can survey nearby suns to see if they have any places that we would like to go and visit.

## Space Arks

The subject of interstellar travel has often been discussed by space enthusiasts, and the asteroids may play an important part in realizing this dream. Typical ideas about interstellar flight fall into five general categories:

1. Scientists who say that interstellar flight is impossible.
2. Those who say that interstellar flight may be accomplished at a fraction of the velocity of light if some form of suspended animation can be achieved.
3. Those who believe that generation after generation could live out their lives on thousand-year-long voyages to the stars in giant spaceships (which might be hollowed-out asteroids).
4. Those who say that the speed of light could be exceeded for space voyagers, but not for the people back home on Earth.
5. And finally, there are those science fiction writers who imagine faster-than-light travel such as that depicted in "Star Trek."

Almost invariably it is assumed that interstellar voyagers will want to return to Earth after their trip. We question this idea. Among our forefathers were many who left the Old World for the colonies without any expectation that they would ever go back "home." The New World *was* their home. In similar fashion, a century later, many people left the East to settle in the West, again without expectation that they would one day return. We do not subscribe to the idea that all adventuresome people who will go to a new frontier just to see what is there are now extinct.

For adventurers such as these, asteroids may play the role of a "space ark" needed to sustain life in the interplanetary void as described in categories two and three above. Such an "ark," created from hollow planetoids, have been suggested by scientists such as L. R. Shepherd, I. M. Levitt, Dan Cole, and J. D. Bernal.[1]

However, the motivation for "century long" interstellar flights in the late twenty-first century and beyond of advanced technological humanity should not be evaluated by the aspirations and values of late twentieth-century technologies. The short-term wants of astrocolonists in a hollowed-out asteroid should leave them reasonably content to spend time building a stable civilization out there before venturing deeper into the cosmos.

There would be no reason for "asteroidnauts" ever to return to Earth, or to colonize new planets, or to send back raw materials. They would be far too advanced technically to consider living on any other solar system planet or moon. They could use the materials and energy resources on their hollowed-out worlds to prepare for a later conquest of the Milky Way. They could regularly broadcast reports of their life and discoveries back to Earth.

## Why Go?

When men can fly one day to the asteroids and the planets—and eventually to the stars—the question arises: Why go? Michael Faraday, the discoverer of electricity, was asked the question: "Of what use is electricity?" He replied with a question of his own: "Of what use is a baby?" We might reply in the same manner when queried about the rationale for crewed colonies in space.

One of the main motivational factors for space colonists will most probably be the desire for new living space away from the overpopulated Earth, as well as to seek new sources of energy, material resources, and knowledge about our origins. With the prospects of the human population of the Earth doubling and even tripling in the next century, these reasons make sense.

The economical advantages of hollowing out asteroids as a primitive, igloo-type, space home compared with constructing giant Earth-made space stations at greater expense bears careful examination. The approach of exploiting and colonizing the asteroids will give mankind the opportunity to cut our collective teeth on peaceful ways of working together with other spacefaring nations.

Capturing and converting asteroids to become humanity's first permanent home in space can go a long way to help us answer the intriguing questions of "Where did we come from?" "Who are we?" and "Where are we going?"—and finally, "Will we meet anyone out there?" in our interstellar voyage in the cosmos.

But, along with these speculations about the future, we need to consider what needs to be done in the immediate future to meet the challenge of the asteroid: Will we use them, or will they destroy us?

## Note

1. All of these scientists are cited in Donald Cox and Dandridge Cole, *Islands in Space: The Challenge of the Planetoids* (Radnor, Penn.: Chilton, 1964), p. 171.

# 21

# Binding the World Together

Together with Europe, Japan, and Canada, the United States and Russia are working to forge a common approach to the exploration and utilization of space. There is much to be learned out there and all of the nations of the Earth have a stake in what we bring back: the seeds of knowledge.[1]

A common theme in science fiction is how all of the peoples of the world unite to battle evil aliens who have come to destroy the Earth. Well, this is no longer science fiction. The difference between fiction and reality, however, is that the aliens that threaten us now are not intelligent; they are cosmic objects that are part of our solar system. Nevertheless, they do offer a challenge for humanity to unite and work together for mutual salvation.

## We Need Insurance!

The dinosaurs did not know they were in danger, and did not have spaceflight. Hence they were powerless to affect their fate. It is very unlikely that our entire species will be threatened with extinction anytime soon, but we know that there are doomsday rocks out there (whether they are asteroids or comets makes little difference) that could impact Earth at any time and destroy a major city by blast and fire, or devastate whole coastlines with tidal waves. We were willing to spend billions to defend ourselves against nuclear attack from a human enemy. We should now be prepared to spend millions to defend ourselves against errant cosmic bodies.

Another way of looking at this is that the probability of an American being killed by an asteroid is about the same as an American being killed in an airplane crash. Those are very good odds, but the insurance companies still make millions of dollars selling airline trip insurance. Of course, for other cultures, the risk of airplane accidents is generally smaller, since most cultures do not travel as much. So, is it worth a dime a year, per person, to defend Earth against such a disaster? If the answer is yes, then the world should be willing to spend hundreds of millions of dollars per year on planetary defenses. At present we are spending almost nothing for this purpose.

We want to emphasize that this is a global problem and that we believe the nations of the world should band together to do this. We do not know when we will be hit by a killer cosmic body, only that, one day *we will*. And it could happen anywhere in the world. Humanity needs to have a global organization to deal with this threat. For openers, it may be okay if the underdeveloped nations of the world are not able to pay their share of the tariff, because we believe that the benefits of the kind of space program that must be mounted to keep the world safe from cosmic threats will benefit the participants enough that they will profit from their altruism. The underdeveloped nations will see this, and join later when they are able.

We also want to stress that this probably needs to be a *new* international body. All of those which presently exist are either so involved in special interest activities, so limited in capability, or so torn by philosophical and ideological wrangling that they are unable to mount a single-minded effort toward our protection. We will return to ideas about this international organization later, but first we will consider the steps already taken in this direction.

First let us consider the United States' response to this mind-numbing possibility.

## Some Positive Fallout from Shoemaker-Levy 9

When the planet Jupiter was hit with twenty-one pieces of the Comet Shoemaker-Levy 9 (S-L 9) in late July 1994, it became the biggest show of cosmic violence in the history of humankind. The death throes of S-L 9 were seen by millions. Even backyard astronomers with small telescopes were able to witness the spectacular hits on Jupiter. But to many observers, including members of the House of Representatives' Science, Space, and Technology Committee, the comet's main impact was not just on Jupiter, but marked a warning shot to the Earth.

On July 20, 1994, while Jupiter was still being bombarded by pieces of S-L 9, producing Earth-sized fireballs, this committee, chaired by George E. Brown, Jr. (D-Calif.), voted to require NASA to track any major comets or asteroids that threaten the Earth. Although astronomers had already cited one hundred objects whose paths occasionally cross the Earth's orbit, they guesstimated that there may be 2,000 more speeding through space that are big enough to do global damage—if they got sucked in by the Earth's gravity.

Initially, the Jupiter fireworks rekindled interest in the slumbering federal plans to create a $300 million Spaceguard early warning network, as advocated by the Morrison committee. (This is similar in intent to the Ballistic Missile Early Warning System that watched for missiles during the Cold War.) Such a network would be a globe-girdling system of high power telescopes. Which could alert the Earth that a stray intruder from elsewhere in our solar system was coming toward us. Then we could mount an effort to deflect this body to save ourselves from a cosmic holocaust.

However, the Representatives were impatient with the twenty-five-year time period to discover all these threats proposed by Morrison. Therefore, they called on NASA to draw up plans for a system that could more quickly detect asteroids and comets that might threaten the Earth. The committee insisted that the task be completed in ten years (by 2004), a rapid pace considering the enormity of the job.

NASA, which needs to find a logical new mission if it is to retain public support, reacted promptly to the committee's vote. It started a feasibility study of a warning system by appointing a six-member panel to be headed by Dr. Eugene Shoemaker, codiscoverer of S-L 9, to define it. This appointment was appropriate, editorialized the *Scientific American*, since Shoemaker had long advocated such an effort.[2]

Congressman Brown, who is also a physicist, stated optimistically that, "It's going to be easy to sell (this early warning network) to Congress" in the wake of the Jupiter explosions.[3] Speaking to a *New York*

*Times* reporter on July 30, Brown drew the analogy with earthquakes, noting that nothing put federal money into geophysics faster than when earthquakes devastated major cities. Brown and his allies in Congress started lobbying for putting the early-warning program into the budget for Fiscal Year 1996 (which began October 1, 1995).

Brown feels that the customary critics of such a program will most probably take a recess from their usual "cosmic carping" as a waste of the taxpayers' money. Dr. Robert Park, a University of Maryland physicist and spokesman for the American Physical Society, the most elite group of this profession in the United States, observed; "Nobody is going to dismiss this possibility. They may have before, but they're hiding now."[4] Now Brown is accepting congratulations from his congressional colleagues for his foresight in preparing to meet this threat from outer space.

Since there has been *no global early warning* comet-asteroid program comparable to the nuclear warning systems which were operational at the height of the Cold War, the scanning of the skies for potential intruders has been reduced to scattered "mom and pop" operations run by a few dedicated scientists, often using borrowed telescopes. (In the case of S-L 9, the team used an eighteen-inch telescope on Mt. Palomar near Pasadena, California, with a bare-bones budget. Shoemaker, a retired geologist; with his wife, Carolyn; and David Levy, an amateur astronomer from Canada, were a prime example of comet-asteroid hunters, searching the skies for needles in the cosmic haystack.)

They agreed it was pure luck when they found their ninth comet together, using partially exposed film to take the historic picture of S-L 9 under less than ideal conditions, which was able to capture their "string of twenty-one pearls," as they dubbed their find. Levy observed in retrospect that the comet might never have been found if the team of three avid astronomers had decided to scrap their observations that night.

The Shoemaker committee delivered its initial recommendations in June 1995 (see chapter 6). NASA promptly refused—when the report was issued—to spend any of its $14 *billion* budget to provide the $4 *million* that the Shoemaker committee declared was needed to begin their early-warning program. It is not clear whether this is blind shortsightedness by NASA, or another Washington political game to hold up a program popular in Congress so as to get something else for the agency.

Some researchers, notably Dr. Edward Teller, the father of the H-bomb, urged that tests be conducted to determine whether missiles armed with nuclear explosives could destroy or deflect objects headed our way. Shoemaker countered that his committee was chartered only "to study detection and not deflection." His personal view was that "it's very premature to consider deflection, because the odds are very low that we will find something that's a real threat."[5]

Shoemaker's analysis of craters on the moon and the Earth suggested that our planet was likely to be struck only once every 100,000 years by an asteroid bigger than one kilometer (half a mile) across, which would be large enough to trigger devastating effects worldwide. He believes that an early-warning system would be a worthwhile investment, noting that if Shoemaker-Levy 9 had struck the Earth instead of Jupiter, it would have precipitated a "global catastrophe." His view is that we should worry about deflection after a threat is found.

So it would appear that at least the U.S. Congress may be ready to acknowledge the problem, even if NASA is not.

## International Steps toward a Joint Crewed Exploration of Mars

Meanwhile, the international community has become interested in joint efforts in outer space even beyond International Space Station Alpha, which is now being built by the United States, Russia, Europe, Canada, and Japan.

In recent years there have been several interesting international Mars exploration proposals floated about for consideration as the next major goal for humanity in the cosmos. One of these was the recommendations of a Stanford University-USSR team of space engineers made in 1991 to define a crewed Mars exploration project.

On June 4, 1992, the team met in Palo Alto, California, to examine the proposition that an international crew of six explorers could be landed on Mars by 2012 and returned to Earth a year and a half later. The concept was based on utilizing the Russian Energia rocket booster, and the estimated cost of this plan (which was to involve additional international participation and lead to a long-term settlement on Mars) is $60 billion over a twenty-year period.

There are four immediate goals of such a practical proposal (which could include a possible subsidiary contact with the asteroid belt):

1. The shortening of the timetable by using an unmanned test of the Mars ascent vehicle to conduct a sample return mission.
2. Using fuel tankers in Earth orbit to extend the payload of the Mars shuttle for the crew and for Mars cargo vehicles.
3. Adding cargo flights to Mars to expand the living space at the initial planetary base, including closed-cycle greenhouse equipment and test machines to use Mars materials.
4. Longer range second and third visits to explore landing sites up to 1,500 kilometers from the primary landing base.

This latest serious Martian space exploration plan is based on existing technologies wherever possible. The Stanford-Russian team believes that our current near-Earth crewed exploration programs can make a smooth transition to a planetary exploration program within the current levels of spending, which could be lessened through international participation.

The timetable for this project, which could have both a direct and indirect impact on asteroid searches in the future, proposed the first test flight of the Mars shuttle in Earth orbit by 2004 or 2006; the return of materials from Mars by 2007; the sending and landing of the first crew to Mars by 2009; and a permanent settlement between 2011 and 2013. This timetable could be accelerated with the discovery of more or new Earth-threatening asteroids.

We consider this to be an interesting step toward increased international cooperation in space, but we also believe that the timetable is too leisurely, and that until we have enough knowledge to define what steps we can and should take in the event we discover a cosmic object on a collision course with planet Earth, the initial focus should be on asteroid exploration. At present, our knowledge of these bodies is so scarce that we might easily make an inappropriate intervention and cause more damage than we prevent.

## International Interest in Asteroid Study

The astronomical program outlined in the Morrison report already has some international support. The report noted

There is a burgeoning awareness in the astronomical community that the NEO impact hazard is a topic that requires attention for reasons other than altruistic scientific pursuit. At the 1991 General Assembly of the International Astronomical Union held August 1, in Buenos Aires, Argentina, the following resolution was passed:

"The XXIst General Assembly of the International Astronomical Union [held in 1991],

"Considering that various studies have shown that the Earth is subject to occasional impacts by minor bodies in the solar system, sometimes with catastrophic results, and

"Noting that there is well-founded evidence that only a very small fraction of NEOs (natural Near-Earth Objects: minor planets, comets and fragments thereof) has actually been discovered and have well-determined orbits,

"Affirms the importance of expanding and sustaining scientific programmes for the discovery, continued surveillance and in-depth physical and theoretical study of potentially hazardous objects, and

"Resolves to establish an ad hoc Joint Working Group on NEOs . . . to:

"1. Assess and quantify the potential threat, in close interaction with other specialists in these fields,

"2. Stimulate the pooling of all appropriate resources in support of relevant national and international programmes,

"3. Act as an international focal point and contribute to the scientific evaluation, and

"4. Report back to the XXIInd General Assembly of the AU in 1994 for possible further action."[6]

Thus it appears that the international science community is also prepared for a cooperative effort to address the problem of devastation by COTEs.

## Need for an International Program

There are three very simple, yet compelling, reasons why human efforts to tackle this problem need to be international in scope.

First, the problem is global in nature. A COTE can strike anywhere on the globe. The odds are that it will strike water, but, as we know, that doesn't reduce the danger. An ocean impact will raise a tidal wave that can travel for thousands of miles, and kill tens of millions of people along the shores, where the world's populations are most concentrated. A strike on land, on the other hand, is more likely to strike a desert than a city, since the former still occupies a much larger proportion of the Earth's land. But, everyone on Earth is at risk. It is the collective business of *all* of us.

Second, we will need early-warning stations all over the Earth to provide a constant watch over the night skies. This will be true even if we institute a space telescope warning system as suggested earlier. We may also need launching sites of several nations to send out interceptors in a timely fashion. At present the world has a scarcity of launch sites, and they are scattered around the globe. We may need them all, if the threat comes soon. Further, no nation or group has a monopoly on the intelligence or imagination needed to deal with a problem of such a magnitude. We need the efforts of everyone to be sure we see, and address, all possible aspects of the problem and its solution.

We would note, at this point, that the scientists, dispassionate lovers of statistics, seem most concerned over large asteroids or comets which threaten our whole civilization. They worry about objects a half-mile wide or larger. Perhaps their statistics do show that the probability of being killed by such a huge object is larger than for the small ones. But consider again 1993 BX3, which was discussed earlier. If this 50-million ton object hit Earth, the impact would be the equal of 50,000 Hiroshima-sized bombs. Maybe that does not scare the scientists with their probability calculations, but it sure scares us! And that is a small asteroid, one that even the Shoemaker team does not predict we will see very long before it hits us. Even that group proposes looking only for objects 1,000 times as deadly. And there are vastly more of the "astronomically insignificant" bodies coming our way than of the large ones.

Third, and most compelling of all, we will probably need nuclear weapons to address the threat when it comes. If the threat comes from a *large* asteroid, we may get enough warning to have the time to deal with it through more benign methods. But by the very nature of long-period comets, we *will not* get enough warning to do anything except use our most powerful tools to dislodge them from their homicidal path. And if it is a small (but very deadly) asteroid, the detection systems proposed will not see it in time to give us any option except nuclear weapons. We do not want to return to the days when we had to go to bed wondering if the world, or much of it, would be destroyed by nuclear weapons before we woke the next day. We already have the nuclear bombs needed for our planetary defense, but almost no one would be willing to give this much power to any single nation or group. These devices *must* be under international control. People of good emotional health and high ethical standards, from all over the planet, with strong emotional attachments to home and loved ones, must be the guardians of this power. Any other approach is just unthinkable.

## The Basis for an International Body

There are already in place two communities that operate on a world-wide, international basis. These are the scientific communities and the world's military forces. Both already operate globally, and the talents of both will be needed in any planetary defense program. We propose that this new global organization be named Spaceguard Command, in honor of Arthur C. Clarke, who first suggested the name in the preface from *Rendezvous with Rama*, quoted at the beginning of this book.

But why, you may ask, do we need a new organization? Why not just use the United Nations? The United Nations has proven itself to be quite ineffective in many situations where a forceful response is needed. There are many examples, the most recent being Bosnia, where the UN proved to be unable to attain a peace or to enforce international sanctions. Sadly, that organization is too bedeviled by religious, ethnic, and cultural quarrels to be able to act clearly and decisively in an urgent situation. We need an organization devoted to the single purpose of planetary defense, one that will not be diverted by nagging questions of economic policy, population control, political ideology, or any other side issue.

The world's military forces are quite accustomed to forming temporary alliances with other military organizations for a specific purpose. The most recent example is the Gulf War, in which soldiers, sailors, and air personnel from countries of widely different cultures banded together to confront an aggressor. In World War II the United States was allied with the USSR against Germany and Japan. A few years later, during the Cold War, the United States was allied with Germany and Japan against the USSR. If military leaders can adjust to rapid policy changes such as these, there is no question that they can work together in an effort that will protect all of their nations at the same time.

Moreover, the military organizations of the world have a scientific and technical capability that is not generally recognized by the scientists of the world. For example, the Morrison committee did not seek out the expertise of the U.S. Department of Defense (DoD) in preparing its recommendations for a sky survey to find threats. When the Shoemaker committee did, they added the assets of the military GEODSS surveillance system to that of existing telescopes and cut the time for discovery of threats from twenty-five years to only ten years, and suggested that this could be accomplished with a much smaller budget!

There is another example. The U.S. military has, for obvious reasons, long had sensors in space to watch for nuclear explosions on Earth. The nature of these systems and their capability has been classified, for equally obvious reasons. Recently, the DoD released some data from their space sensors about the frequency of asteroid hits upon the Earth. The data show hits caused by very small meteors, those causing an explosion of only a thousand tons of TNT or so. These look very similar to nuclear explosions but due to their small size, they dissipate harmlessly, high in the atmosphere.

The first of these data was released in 1993 by Col. Simon P. Worden, who has been involved with Star Wars projects for a number of years. He described one unknown COTE explosion as producing a ten-kiloton (20 million pounds of TNT) impact over the central Pacific. In view of the Gulf crisis in the Middle East, "it could have been mistaken for a nuclear detonation and could have triggered very serious consequences."[7] It was recognized for what it was, and no harm was done.

More recently, in October 1993, a more complete report was issued by the DoD, showing that there have been an average of eight asteroid hits a year, each a kiloton or larger, in the interval between 1975 and mid-1993.[8] A total of 136 such hits were recorded, and it is possible that some were missed. These data are beginning to convince astronomers and other scientists that the frequency of meteor hits upon the Earth is much greater than they had previously believed.

The other element needed in a Spaceguard Command organization is scientific expertise. Scientists are already used to working together in a variety of formal and informal ways. You may remember that it was under the auspices of the International Geophysical Year that the first Earth satellites were launched; first by the USSR and then by the United States.

There are many good examples of international groups that operate to exchange information. One example is the International Astronautics Federation, which meets annually to exchange data about various aspects of space flight. Both Russia and the United states have long participated in this exchange, even at the height of the Cold War. The exchange was limited and guarded in those days, and many people with sensitive duties were denied an opportunity to participate in these exchanges. However, the precedent was set for the exchange of scientific data, even under very adverse circumstances.

Another example of an international body having an important governing function is the World Radio Conference. This group meets periodically and decides what radio services may use what parts of the spectrum, and in what parts of the world. This is a very controversial and challenging task. There is always demand for more radio spectrum than is available, and various nations and interest groups conduct high powered lobbying and political campaigns to have their desires fulfilled. This has been going on since the early days of radio, and the decisions made by this international body have always been globally respected; even during time of war, the belligerents respect the rulings of this body. This proved that an international body devoted to just one aspect of international affairs can operate successfully even under the most trying of circumstances.

## A Specific Proposal

For these reasons, we favor a *new* international organization to address the threat of cosmic impacts to our planet. This body would be composed primarily of scientists and military people working together toward one very simple goal: preventing damage to the Earth (and its inhabitants) from a comet or asteroid impact.

The formation of the National Aeronautics and Space Agency (NASA) provides a useful precedent for beginning the new international organization. NASA was formed in 1958 to answer the *Sputnik* challenge of October 4, 1957. To form the new agency various elements of the venerable National Advisory Committee for Aeronautics, including Langley Research Center in Virginia, the Lewis Flight Research Center of Ohio, and the Ames Research Center in California; the Jet Propulsion Laboratory of Cal Tech in California; and large portions the Army Ballistic Missile Agency (ABMA) in Alabama and the Air Force missile test range at Cape Canaveral in Florida were integrated. There are already in existence a variety of international organizations that could be combined in a similar fashion to form Spaceguard Command.

We describe this new agency as Spaceguard Command because we envision a small group of high ranking scientists and military officers performing the executive and policy functions of this organization. Most of the time, this group would simply coordinate the exchange of information. If there is no threat to the planet in sight, this would be their primary function. Such an exchange could be implemented via international conferences, the electronic networks (such as the internet), and educational programs to keep the world informed of what is happening in the heavens.

This policy group would campaign to bring all of the world's information-gathering agencies, and their instruments, into the information exchange. This will be a challenge, because many of the world's most capable military/intelligence systems are kept under very tight security. The most guarded secret a nation has is the answer to "how good is our intelligence?"

A case in point is the National Reconnaissance Office (NRO) of the United States, whose very existence was not acknowledged until that fact was declassified in 1992; it had been just a shadowy rumor before that. Most of what it does, and its budget and capabilities, are still highly classified. It is known that this is the agency which operates spy satellites to watch what other nations are doing on Earth. It has earned high praise from several Congressmen for the performance of its global satellite coverage. It is presumed to be at the very forefront in imaging technology. It seems reasonable to ask if these spy satellites can be turned around to look up for stray cosmic bodies when they do not have urgent business looking down at Earth. If this could be done, our spy satellites could act as mini-Hubbles looking for COTEs. This would be a mission added to Vice President Gore's recent call for the NRO to do ocean surveillance, study the expansion of deserts, and check on floods and water pollution.

We cannot answer the question of whether this plan is feasible, but the military establishment could. This is another reason why involving them in any international planetary defense is essential.

In any case, it is clear that the technology of these satellites would be applicable to building space telescopes such as those proposed earlier. Moreover, it is clear that the amount needed to build a new $50 million set of telescopes, as proposed by the Morrison committee, or the $4 million requested by the Shoemaker committee, would be pocket change in the NRO's annual budget.

A reconstituted NRO could be merged with various government agencies such as the National Security Agency, the National Oceanographic and Atmospheric Agency, the National Imagery and Mapping Agency, and other presently decentralized agencies to form the bedrock for Spaceguard Command—constructing the new organization on a solid base modeled after the birth of NASA. In this way, Vice President Al Gore's dream of recycling the NRO to have a new double mission looking *up* into the cosmos to search for the asteroids threatening us as well as *down* at the Earth could come into being.

We presume that other nations also have military systems, such as the GEODSS and spy satellites we have already mentioned, that would be as useful to a planetary defense. With tensions between the superpowers as low as they have ever been, now is as good a time as we are ever likely to find to urge them to share information for our planetary defense.

There is also an important civilian side to this Spaceguard Command. We have described the *NEAR* spacecraft, now on its way to the asteroid Eros. We need many more more missions like this, and this should also be an international effort. With over seventy predicted close approaches in the next twenty-five years, there is no shortage of targets for such exploration. The space agencies must exchange data on the most effective designs to use for the spacecraft and the launchers. The scientists of all nations need to come together, as they do now, but with greater urgency, to share their data about the space surrounding Earth.

A Spaceguard Command would accumulate and store, in the greatest protective custody, the vast arsenal of nuclear weapons made surplus by the emergence of peace. These weapons would be fitted to a few spacecraft designed to replace commercial satellites for launch purposes, as described earlier. Once these weapons and their carrier satellites were safely stored, the Spaceguard Command would have little to do except look through its telescopes, space- and Earth-based, seeking any threat to our planet. When one is detected, it would then preempt as many space launches as it required to deal with the emergency. Of course, the procedures for such an action need to be established by treaty and international law long before such an emergency arises.

In time, these weapons may become obsolete. More likely, the carrier spacecraft will become outdated first, and will need to be replaced by more modern versions. But the annual budget for such an activity will be minuscule compared to the global peace of mind we will enjoy, knowing that there is a response for the day, hopefully remote in time, when we detect an incoming cosmic object. At some more future time, the weapons in this Spaceguard Command arsenal may be called upon to move asteroids around to benefit mankind. Since such an activity is very dangerous if done with malice, it is again most appropriate that this be done by our international guardians.

## A Dire Prophecy for the Planet Earth

David Levy, the amateur astronomer from Canada who helped discover Comet Shoemaker-Levy 9, has described what would happen if a similar size comet or asteroid hit the Earth. It could possibly have the following results: Dig a crater fifty miles wide and ten miles deep, no matter where it hit, destroying the protective ozone layer in the bargain. The temperature would soar above 212 degrees and darkness would envelope most of the Earth as a dust cloud blotted out the sun. A fireball on contact could soar as high as 2,000 miles and 75 percent of all species of life forms would be obliterated, as were the dinosaurs at the end of the Cretaceous Period of the Earth's history.

Unlike the dinosaurs, we have a choice not to be made extinct by such a cosmic event. The question for our species is whether we have the intelligence, the imagination, and the will to do something before it is too late. And, of course, we do not know when too late will be, so we need to start now. The positive side is that if we choose, we can also have a space exploration and colonization effort that will make the wonders of today appear primitive.

As the American Institute of Aeronautics and Astronautics recently concluded in its September 1995 position paper, "If some day a [large] asteroid does strike the Earth . . . and we could have prevented it but did not . . . then it will be the *greatest abdication of responsibility in all of human history.* [Emphasis added.]"[9]

## Notes

1. Albert Gore, "Spacebeat," *Ad Astra* (October/November 1995): 8.

2. "Science and the Citizen" (editorial), *Scientific American* 271, no. 4 (October 1994): 16–20.

3. *New York Times,* 31 July 1994, p. C1.

4. Ibid.

5. *New York Times,* July 8, 1995, p. C4.

6. *Report of the NASA International Near-Earth-Object Detection Workshop* (Washington, D.C.: GPO, 1992).

7. Cited in *Proceedings of the Near-Earth-Object Interception Workshop,* G. H. Canavan, J. C. Solen, and J. D. G. Rather, eds. (Washington, D.C.: GPO, 1993), p. 206.

8. J. Kelly Beatty, "Impact Secrets Revealed," *Sky & Telescope* 87, no. 2 (February 1994): 26–27.

9. American Institute of Aeronautics and Astronautics, September 1995 position paper.

# Epilogue

Several events occurring since the completion of the main manuscript add even more emphasis to some points that we made earlier. We will briefly summarize these here.

On May 14, 1996, it was announced in the *New York Times* that a new joint U.S. Air Force and NASA program to detect close approach asteroids had been operating since December 1995. A one-meter (thirty-nine-inch) telescope on ten-thousand-foot Mt. Haleakala, Hawaii, went into operation then, searching for COTEs. This joint effort is called Project NEAT (for Near Earth Asteroid Tracking). The principle investigator for the project comes is Dr. Eleanor Helin of the Jet Propulsion Laboratory, a noted asteroid hunter.

The first few months of operation were disappointing, since there was almost continuous cloud cover over the telescope at night. But the sky cleared, and they began to discover many new asteroids each month. In the first few months they found several previously unknown objects that are a potential threat to Earth. The largest of these new threats, designated 1996 EN, is 3 kilometers (1.8 miles) in diameter, quite sufficient to wreak major havoc upon Earth if an impact occurred. By the end of July 1996 2,239 new objects had been spotted and enough orbital definition was obtained to assign 333 new designations.*

Sadly, the government has been unwilling to provide an additional $50

---

*Readers who have access to the internet can find the latest details on this exciting search on the World Wide Web at http://huey.jpl.nasa.gov/~spravdo/ neat.html.

297

million which has been requested to augment this search for cosmic threats. Specifically, NASA has been unwilling to add this to its budget request to Congress for asteroid searches, although it has provided a very small amount of funding for Space Watch (run by Tom Gehrels) and for NEAT.

The danger of this attitude has been demonstrated very recently. On Sunday May 19, 1996, a newly discovered COTE (1996 JA1) whizzed past the Earth at a speed of 36,000 mph. It was discovered just four days prior to that by some college astronomy students looking through a telescope atop Mt. Lemmon, near Tucson, Arizona. It missed us by only 279,000 miles (in other words, it passed closer to Earth than the moon does). One astronomer termed this "a close call."

We want to point out that an asteroid need not be aimed directly at Earth to cause a problem. The Earth's gravity field will "bend" a COTE's trajectory slightly as it gets close, so that a "near miss" may, in fact, result in a collision. In other words, the "capture area" of the Earth is several times larger than its actual area.

Asteroid 1996 JA1 was estimated to be a third of a mile in diameter. This would have caused an explosion nearly a hundred times as powerful as the one that hit Tunguska in 1908. And that, you may remember, caused devastation over a vast area of Siberia. This recent intruder marked the closest threat to Earth in recent years, and should be considered by our fragile planet as a warning shot about future life-challenging visitors whirling around in our solar system.

Less than a week later, on May 25, 1996, the world got a second cosmic scare when asteroid JG passed within 1.9 million miles of Earth. This one was even larger, three-quarters of a mile in diameter. According to the estimates made by scientists, this is about the size that would cause a planetary disaster; at least a *billion* people would be killed, and modern civilization would be destroyed. Furthermore, it had been discovered only two weeks before its close approach, on March 8, by an Australian astronomer, Robert McNaught of the Siding Spring Observatory of New South Wales.

These two back-to-back close calls dramatize, in our minds, the need for an international asteroid detecting organization. But there is still no call to action from Congress or the White House.

The good news is that the U.S. Space Command has been studying this problem of planetary defense. Col. Pete Worden, who directed a project in mid-1996 that planned to send *Clementine* to the moon to return thousands of new photographs, is now with the requirements directorate at Space Command in Colorado. In a recent *Space News* article, Col. Worden said "Throughout the Air Force there is strong interest in [organizing planetary defense against cosmic objects.] But at the present time it is not an assigned mission. My personal feeling is that it will be."[1]

John Darrah, chief scientist for the Space Command, told *Space News* that objects much smaller that the one-kilometer size cited by astronomers (the ones we have been much concerned with in earlier chapters) are a hazard to Earth. The small ones are much more numerous, and much harder to spot, so that they can much more easily hit us with little warning. There is a good chance that in a decade or two will will know about all of the "big ones" that might hurt us, and be able to predict their advent decades to centuries before they strike.

This is certainly a step in the direction we advocate in chapter 21. One of the four main components of a planetary defense (the U.S. military) seems willing to get involved. The other three, foreign military, and domestic and foreign science organizations have yet respond, but we think prospects for a planetary defense have taken an important step forward.

It is worthy of note, however, that China, in refusing to sign a comprehensive nuclear ban treaty, cited the danger of asteroids as one reason they wished to maintain a nuclear arsenal. Russian President Boris Yeltsin visited Chinese President Jiang Zemin to seek agreement to a comprehensive Nuclear Test Ban Treaty. Zemin cited the asteroid threat and the need for "peaceful nuclear explosions to combat it" as reason not to sign the treaty at this time. Zemin did tell Yeltsin that he was in favor of concluding a test ban treaty, but not now. Perhaps at least one nuclear and space power has recognized the asteroid threat and will be amenable to an international planetary defense solution along the lines we have proposed.

To conclude on a happy note, we want to describe the prospect of finding diamonds in asteroid impact craters. In the late spring of 1996, the American Geophysical Union, the world's largest professional group devoted to Earth studies, held its first meeting ever on impact diamonds (which are formed when a meteor strikes the Earth) and mineralogy. Organized by Dr. Peter Fiske, a planetary scientist at the Lawrence Livermore Laboratory in California, the conference concluded that the rocks of old craters formed by comet and asteroid impacts probably hide thousands, if not millions, of tons of diamonds!

This startling discovery, according to the experts, is likely to be more of a boon for understanding Earth's cataclysmic past that for a new wave of worldwide commercial mining. (Some of the diamonds are only the size of a matchstick head, while others are peanut size.)

What is so fascinating about this latest asteroid hypothesis is that with the known 150 authenticated impact craters (with an added five new ones discovered every year) the geologists are finding new evidence of past collision fury.

Consider some of the world's largest known craters:

- Sudbury, Ontario, Canada, which left a giant 155-mile-wide scar across the terrain 100 million years ago;
- Popigai crater in the wilds of northern Siberia, 60 miles wide and 35 million years old; or
- Chicxulub crater in the Yucatan, 150 miles wide and 65 million years old which is now believed to be the site of the impact that killed the dinosaurs and made way for mammals.

We may be on the brink of a global diamond hunt like nothing the world has even seen before. These prime targets, wrought by the high temperature and pressure of impacts on carbon to create diamonds, could open new windows on the mysteries of creation as well as a rich mother-lode of new diamond sources.

Interestingly, it was the Russians who first discovered that shock metamorphism could transform the carbon of bedrock into diamonds and that the asteroid diamonds are larger and sometimes easier to see with the naked eye than those found in meteorites. This discovery, made in the early 1970s, was kept a state secret for nearly two decades until the collapse of the USSR, when the information began to leak out, including the fact that the Russians had investigated mining such impact craters for diamonds.

Now it is a new ball game for exploratory mining of the world's impact craters, thanks to our cosmic neighbors, the asteroids.

## Note

1. Leonard David, "U.S. Air Force to Study Asteroid Defense Plans," *Space News* (July 15–21, 1996), p. 17.

# Appendix A

# Selected Bibliography

Alverez, L. W., et. al. "Extraterrestrial Cause for the Cretaceous–Tertiary Extinction. *Science* 208 (1980): 1095–1108.

Beck, Melinda, and David Glick. "Doomsday Science." *Newsweek,* 23 November 1992, 56–61.

Binzel, Richard, et al. *Asteroids I.* Tucson: University of Arizona Press, 1979.

———. *Asteroids II.* Tucson: University of Arizona Press, 1990.

Broad, William. "Asteroids a Menace to Earth Life: Could Destroy the Earth." *New York Times,* 18 June 1991, C1 and C7.

———. "Asteroid Defense: 'Risk is Real' Planners Say." *New York Times,* 7 April 1992, C1 and C7.

Chapman, C. R., and D. Morrison. *Cosmic Catastrophes.* New York: Plenum, 1989.

Clarke, Arthur C. *The Hammer of God.* New York: Putnam, 1992.

———. *Rendezvous with Rama.* New York: Ballantine, 1973.

Cox, Donald, and Dandridge Cole. *Islands in Space: The Challenge of the Planetoids.* Radnor, Penn.: Chilton, 1964.

*Doomsday Asteroid* (videotape). PBS-TV/WGBH, Boston. 1995.

Helin, E. F., and R. S. Dunbar. "International Near-Earth Asteroid Search." *Lunar and Planetary Science* 15 (1980): 358.

Morrison, David. *The Spaceguard Survey: A Report of NASA's Near-Earth Object Detection Workshop.* Washington, D.C.: Office of Space Science and Applied Solar System Exploration, 1992.

301

Pike, J. "The Sky Is Falling: The Hazard of Near-Earth Asteroids." *Planetary Report* 11 (1992): 16–19.

Rabinowitz, D. L. "Detection of Earth-Approaching Asteroids in Near-Real Time." *Astronomical Journal* 101 (1993): 1518–29.

Raup, David. *Extinction: Bad Genes or Bad Luck?* New York: W. W. Norton, 1992.

Remberger, Boyce. "Death of the Dinosaurs." *Science Digest* (May 1986): 28+.

Shoemaker, E. M., and R. F. Wolfe. "Asteroid and Comet Bombardment of the Earth." *Annual Review of Earth and Planetary Sciences* 11 (1983): 461-94.

Steel, Duncan. *Rogue Asteroids and Doomsday Comets.* New York: John Wiley & Sons, 1995.

Trefel, Jane. "Stop to Consider the Stones That Fall from the Sky." *Smithsonian* (September 1989): 81+

# Appendix B

# Glossary

**Absolute Magnitude (of a planetoid).** This is the magnitude a reflective object (e.g., asteroid) would have when it faced the sun and was one AU from Earth and from the sun.

**Acquisition.** The process of locating an asteroid (or other object) that is to be tracked.

**Amor Asteroid.** An asteroid that presently stays outside the orbit of Earth (i.e., its orbit is greater than 1.017 AU) but does come within 1.3 AU of the sun. It is not a threat now, but may be in the future.

**Antipode.** Opposite; usually used to refer to an approximately opposite point on a sphere; i.e., Australia is at the antipode of England.

**Aperture.** The diameter of the primary optic of a telescope; a measure of light gathering power, and (for a single optic) of resolving power.

**Aphelion.** The point in the elliptical orbit of a planet, comet, or asteroid that is most distant from the sun.

**Apogee.** The point at which an orbiting object is farthest from the body being orbited.

**Apollo Asteroid.** An asteroid that currently crosses Earth orbit with an aphelion greater than Earth's. A collision with Earth is possible.

**Armature.** The moving part of an electric motor.

**Artificial Gravity.** Synthetic gravity (a centrifugal force) produced by rotation. Often suggested to make space travel more convenient.

**Asteroid.** An object smaller than a planet orbiting the sun. It shows no sign of atmosphere or other cometlike activity. Most asteroids are in the "main belt" between Mars and Jupiter.

**Astronomical Unit (AU).** The average distance from Earth to the sun; about 93 million miles. A yardstick for planetary measurements.

**Aten Asteroid.** An asteroid with a perihelion much closer to the sun than Earth, but one that crosses Earth's orbit. A collision with Earth is possible.

**Baseline.** The data serving as a basis for calculating or locating something.

**Big Four.** The name given to the first four asteroids discovered in the nineteenth century (i.e., Ceres, Juno, Pallas, and Vesta).

**Capture.** The process by which an asteroid is moved from its natural orbit; especially into orbit around Earth. Originally, to capture an asteroid meant to discover its orbit precisely enough that its position at a later time could be predicted.

**Carbonaceous.** A class of chondrite meteorite that contains hydrocarbon materials and (usually) water.

**Charge Coupled Device (CCD).** An electronic equivalent to film, as used in a videocamera or telescope.

**Cheap Access To Space (CATS).** A goal for many space activists, meaning achieving costs to orbit at *least* ten times cheaper than 1996 costs.

**Chondrite.** A stony meteorite composed of finely crystallized material.

**Coil Gun.** A form of electric catapult in which the electricity is passed through coils (or solenoids) to produce the accelerating force.

**Coma.** A roughly spherical region of diffuse gas which surrounds the nucleus of a comet. Together, the coma and the nucleus form the comet's head.

**Comet.** A celestial body rich in volatile matter (ices). As it approaches the sun some of its matter boils away, forming a coma (or head) and (usually) a long tail. It is on a very elongated elliptical path, often approaching a parabola.

**Commutator.** A device, as on a generator or motor, for changing the direction of electric current.

**Composite.** A structural material made of separate parts which together work better than either alone. Examples would include reinforced concrete and fiberglass-epoxy.

**Coriolis Force.** An acceleration produced as a by-product of rotation used to create artificial gravity. It will cause something dropped not to fall straight down, but also go sideways. Large amounts of it will disturb the sense of balance and cause motion sickness.

**Cosmic Object Threatening Earth (COTE).** The term we use to describe any comet or asteroid that may hit Earth. Astronomers usually call these NEOs (Near Earth Objects), but even a month before they would hit Earth they are tens of millions of miles away.

**Cosmogony.** Theory of the origin and development of a solar system.

**Cosmology.** The branch of astronomy that deals with the overall structure and evolution of the universe.

**Crater.** An approximately circular depression, typically surrounded by a raised rim, caused by the impact of a meteorite.

**Declination.** The angular distance north or south of the celestial equator.

**Delta-V.** The measure of performance of a launch vehicle stage; how much velocity it can add to the stages and payload above it. Measured in feet per second, kilometers per second, or similar units.

**Earth-Crossing Asteroid (ECA).** An asteroid whose orbit crosses Earth's path, or may later because of the gravitational influence of other planets, especially Jupiter.

**Earth-Crossing Comet (ECC).** A comet that will cross the path of the Earth.

**Eccentricity (of Ellipse).** A measure of the amount by which an ellipse departs from a pure circle. A circle has an eccentricity of zero and a parabola (which is not a closed figure) has an eccentricity of one.

**Ecliptic.** The plane which contains the orbit of the Earth. Declination is measured from this plane.

**Electric Catapult.** A device used to accelerate something using electricity as the motive power. Catapults on aircraft carriers now use steam, but there are plans to replace them in the future with electric catapults.

**External Tank (ET).** The large tank on the side of a space shuttle (when ready for launch) that carries the propellants.

**Fall.** A meteorite that was seen to fall, and then found on the ground. These are considered to be more representative of asteroids.

**Find.** A meteorite that has been identified as such after it has been found by someone; it was not seen to fall. These tend to be iron objects, hence they are picked up because they are so "unusual."

**Gamma Radiation.** A form of radiation similar to X-rays, but of shorter wavelength; more penetrating than X-rays.

**Gee.** The acceleration of gravity at the surface of the Earth; this causes a pound of mass to exert a force of one pound. In space this relationship is not true, so we must specify the gee level.

**Geosynchronous Orbit (GEO).** An orbit about Earth at such an altitude that it appears stationary to an observer on the surface. Used by many communication satellites.

**Ground-based Electro-Optical Deep Space Surveillance System. (GEODSS).** Developed during the Cold War, this system was designed to protect the United States from unfriendly spacecraft.

**Heavy Lift Launch Vehicle (HLLV).** A space launcher larger than any presently in use. Proposed, but not built.

**High Energy Earth Orbit (HEEO).** An orbit whose apogee reaches out toward the moon. Such an orbit will is difficult to reach from Earth, but may make interplanetary travel much easier.

**ICE.** *I*nternational *C*ometary *E*xplorer.

**Ices.** Volatile material in frozen form, such as water ice. As used in describing comets, it includes other vapors in frozen form, such as dry ice (frozen $CO_2$) and many more frozen vapors.

**IMO.** *I*nternational *M*eteor *O*rganization.

**Inclination.** *Solar orbit:* The angle between the orbit of a comet or asteroid (or other body) and the ecliptic plane. *Earth orbit:* The angle between the orbit of a spacecraft and the equator of Earth.

**Intermediate-Period Comet.** Comet with a period of between twenty and two hundred years.

**Ion Engine.** A form of jet propulsion device that uses electric power and ionized gases to produce thrust. After the gas working fluid is ionized, magnetic fields accelerate it to very high velocity. This produces high specific impulse, but low thrust.

**Infrared (IR).** Infrared light. Radiation with a wavelength too long to be seen by humans, but perceived as "radiant heat."

**Iridium.** A platinum group metal, rare on Earth's surface. It can be identified in very minute amounts by radiochemistry techniques.

**Kevlar®.** A family of plastics (polymers) from DuPont that is exceptionally strong. Used in automobile tires and in engineering plastic materials.

**Kiloton.** The energy equivalent of 1,000 tons of TNT. This is $4.3 \times 10^{12}$ Joules, or 1.2 million kilowatt hours.

**Kinetic Energy.** The energy of motion. It is proportional to the mass of an object times its velocity squared. (Numerically this product is divided by two.)

**Ladar.** The use of laser light in place of radio waves to locate an object and determine its distance and velocity. (See also **Radar.**)

**Laser.** *L*ight *A*mplification by *S*timulated *E*mission of *R*adiation. A device for producing intense beams of light entirely of one color (or wavelength).

**Linear Motor.** Any form of electric motor in which the moving part (armature) travels in a linear rather than circular motion.

**Long-Period Comet.** Comet with a period greater than two hundred years.

**Low Earth Orbit (LEO).** Generally, any Earth orbit that is lower than the radiation belts.

**Magnetometer.** A scientific instrument used to measure the strength of a magnetic field.

**Magnitude.** A logarithmic measure of the brightness of a celestial object. A difference of 5 in magnitude is a brightness difference of 100, with bright objects having lower numbers. (The sun is so bright that its magnitude is a negative number, $-26.7$.).

**Main-Belt Asteroid.** Asteroids that occupy the main asteroid belt between Mars and Jupiter. Sometimes this usage is limited to the densest part of the belt; between 2.2 and 3.3 AU from the sun.

**Mass Fraction.** The ratio of mass of a fueled rocket to its mass with no fuel. It is one of the two most critical indicators of the performance of a rocket. The other is specific impulse, or Isp.

**Megaton.** Energy equivalent of one million tons of TNT ($4.3 \times 10^{15}$ Joules).

**Meteor.** The light streak produced by a small celestial body as it burns up in the atmosphere; often used to mean the object itself. If it is large it is often called a fireball.

**Meteorite.** A natural object of extraterrestrial origin that survives its passage through the atmosphere and hits the Earth.

**Micro-.** A prefix meaning one millionth of. For example, a microradian is one-millionth of a radian.

**Minor Planet.** Another name for an asteroid.

**NEAR.** *N*ear *E*arth *A*steroid *R*endezvous, a space probe, launched in February 1996, planned to conduct the first survey of an asteroid from orbit around it.

**Near-Earth Object (NEO).** Objects whose orbits bring then relatively close to Earth in astronomical terms. Specifically, Apollo, Amor, and Aten asteroids, and some comets.

**NEAT.** *N*ear *E*arth *A*steroid *T*racking project; a joint USAF-NASA search for cosmic objects that may crash into Earth.

**Neutron Activation.** A technique of radiochemistry in which a sample is subjected to neutron radiation to activate isotopes which can then be detected by a radiation counter.

**Neutron Bomb.** A form of atomic bomb which has been designed to produce more neutrons than a normal atom bomb.

**New Comet.** A comet on its first approach to the sun.

**Occultation.** The hiding of one object in the sky (such as a star) by an object closer to the observer.

**Oort Cloud Comet.** A comet originating in a postulated very large, spherical cloud of comets, far from the sun. Perturbations of this cloud send new comets into the inner solar system.

**Opposition.** The case when the Earth, sun, and a planet (or other object) form a straight line.

**Orbit.** The path of body moving in the gravitational influence of a larger body. Examples are a satellite around the Earth, or the Earth around the sun.

**Orbital Period.** The time required for a celestial object (Earth or the moon, for example) to complete one trip (orbit) around its primary (sun or Earth, respectively).

**Orbital Plane.** The plane in which the orbit of any celestial bodies lie.

**Outgassing.** The evolution of gas (or vapor) into a vacuum from a spaceship or a process conducted into vacuum. Very much of this will "spoil" the vacuum, since it will now be contaminated with gases.

**Paleontology.** The study of life forms from earlier geological epochs based upon their fossil remains.

**Parabolic Orbit.** This is an orbit that can just escape from the body in question. The term is usually used to refer to long period comets that are so close to solar escape velocity that it is not certain if they come from the Oort cloud or from interstellar space.

**Paradigm.** A model or system of thought. A new paradigm is more than just a new idea, it is a whole new set of ideas or concepts that fit together.

**Perigee.** The point at which an orbiting object is nearest the body being orbited.

**Perihelion.** The place in the orbit of an object in orbit around the sun where it is closest to the sun.

**Perturbation.** Action causing an orbit to vary slightly from its previous path; generally this is caused by the gravitational field of a planet (or moon), but smaller effects result from solar pressure and other forces.

**Phase Angle.** The phase of a celestial body; a new moon, for example is a phase angle of zero, and a half moon is either 90 or 270 degree phase angle.

**Planetoid.** Another name for an asteroid or minor planet.

**Potential Energy.** The energy of position. It is the product of the mass of an object times its gravitational height. (Note that as gravity decreases with altitude, gravitational height is numerically less than physical height.)

**Propellants.** The chemicals used in a rocket motor; generally one fuel and one oxidizer. In an ordinary chemical rocket the propellants also become the working fluid.

**Protoplanet.** A planet in the process of formation; the term is especially used early in the life of a solar system.

**Radar.** *RA*dio *D*etection *A*nd *R*anging. Use of radio waves to locate an object and determine its distance and velocity.

**Radian.** Natural angular measure. The length of a one radian arc is the same as its radius. Numerically it is about 57.3 degrees.

**Rail Gun.** A form of electric catapult in which the electricity travels from one rail, across a gap and back along a parallel rail to produce the accelerating force.

**Regolith.** The surface layer of a planet that has been affected by weather, or—on an airless body like the moon—by space radiation and micrometeorite bombardment.

**Rendezvous.** The meeting of two or more objects (natural or artificial) at a predetermined time and place.

**Resonance.** An orbital condition in which one object (such as an asteroid) is subject periodically to the same gravitational perturbations; usually referring to Jupiter and one or more asteroids.

**Retrograde Motion.** Orbital motion of a solar system object (such as a comet) which is opposite the motion of all of the planets. In Earth orbit, an orbit with a westward component.

**Right Ascension.** The hour angle of a celestial body, measured east in the ecliptic plane from the first point of Aries, also known as the vernal equinox.

**Roche's Limit.** The closest distance at which a natural satellite can orbit its primary without being broken apart by gravitational tides. (Often described as about two and a half times the radius of the primary body.)

**Schmidt Telescope.** A type of astronomical telescope with a particularly wide field of view; very suitable for searching for celestial objects or satellites. Usually used as a camera.

**Semimajor Axis (of Orbit).** Half of the longest axis of an ellipse. For a body in orbit around the sun, it is the average distance from the sun.

**Short-Period Comet.** Comet with a period of less than twenty years.

**Solid Rocket Booster (SRB).** One of the solid rockets strapped to the shuttle (or other launch vehicle) at lift-off.

**Spaceguard Survey.** A proposal (from a NASA committee) for an international search of the sky for asteroids and comets that pose a threat to Earth.

**Specific Impulse (Isp).** A measure of the effectiveness of a set of rocket propellants. In English units, propellants with a specific impulse of 400 seconds will produce 400 pounds of force of thrust for every mass pound of propellants burned per second. (See also **Mass Fraction.**)

**Space Shuttle Main Engine (SSME).** The largest and most powerful rocket engine in use in the United States today.

**Single Stage To Orbit (SSTO).** A concept for a fully reusable space launch vehicle which is fully contained in one vehicle or "stage."

**Station Keeping.** The process of firing jets to maintain the orbit of a spacecraft on the desired path.

**Tether.** A flexible connecter joining two bodies in orbit, in yo-yo string fashion.

**Tonne.** A metric ton; 1,000 kilograms or about 2,205 mass pounds.

**Triangulation.** A means of determining the distance to a far-away object. The angles to the object are measured at both ends of a baseline, and trigonometry is used to calculate the distance.

**Trojan Point.** A point in the gravity system of two objects (such as the sun and Jupiter, or Earth and moon) where the gravity effects are so counterbalanced that a body can remain there indefinitely. These were first theoretically described by Comte Joseph-Louis Lagrange in 1772. There are five such points for any pair of bodies. The one known as L-5 is often suggested as a location for an artificial space colony. These are called Trojan points after asteroids were discovered at the L-4 and L-5 points in the Jupiter/sun system.

**Trojans.** Two groups of asteroids discovered early in the twentieth century, at the sun/Jupiter Trojan points 4 and 5. These asteroids were named after Greek Trojan war heroes.

**Two Stage To Orbit (TSTO).** Generally used to describe a reusable launch vehicle concept in which two stages are used to attain low Earth orbit, as opposed to SSTO.

**UFO.** *U*nidentified *F*lying *O*bject.

**working fluid.** The material ejected from a jet propulsion device (such as a rocket motor) to cause a reaction force. For a chemical rocket this is the products of combustion of the propellants; for rockets with other energy sources it may be any fluid or solid.

# Appendix C

# Executive Summary
# of the Shoemaker Report

This summary is quoted from the June 1995 NASA *Report of the Near-Earth Objects Survey Working Group*, Chaired by Eugene M. Shoemaker of Lowell Observatory.* The workshop was convened by NASA, Solar System Exploration Division, at the insistence of Congress, after the impact of comet Shoemaker-Levy on Jupiter in July 1994.

Approximately 2000 near-Earth objects (NEOs) larger than one kilometer in diameter revolve around the sun on short-period orbits that can occasionally intersect the orbit of the earth. Only about 7 percent of this population has been discovered. There is about one chance in a thousand that one of these undiscovered objects is destined to collide with the Earth during the lifespan of the average American. Such a collision has the potential of injecting sufficient dark material into the atmosphere to cause a major loss of global crop production and consequent loss of human life.

NASA's charge to the NEO Survey Working Group was to develop a program plan to discover, characterize and catalog, within ten years (to the extent practicable), the potentially threatening comets and asteroids larger than one kilometer in diameter.

---

*(Washington, D.C.: NASA Space Explorations Division, Office of Space Sciences, 1995).

Advances in the last few years in the development of charge-coupled devices (CCDs) have led to substantial improvement in the projected capability to carry out a systematic survey of NEOs. In particular, large format, high quantum efficiency, fast readout CCDs have been developed under U.S. Air Force-sponsored research. The efficiency of these detectors is now close to the theoretical limit. Use of these detectors on sufficiently large telescopes would enable rapid progress to be made in an NEO survey.

We have defined a program that responds immediately to the challenge of discovering potentially threatening NEOs. It would carry out a census of short-period comets and asteroids larger than one kilometer in diameter and seriously address smaller NEOs and long-period comets, as well as develop a broad database of physical observations in order to evaluate the impact hazard. This program, based on further development of existing efforts within the civilian astronomical community, could meet restricted objectives (surveying the short-period NEO population) in ten years following an initial three-year development phase.

In order to proceed promptly, maximum use needs to be made of existing telescopes. In particular:

- we encourage collaboration of the U.S. Air Force. The Air Force facilities and technologies will enhance the undertaking. The Air Force's continued development of large array imaging cameras will be of value to all the participants;
- we encourage collaboration of the international community, including further development of programs underway in France, Australia, China, Canada, Russia, and for the European Southern Observatory.

**Recommended Program**

The recommended program will accomplish the objective of discovering 60–70 percent of short-period NEO's larger than one kilometer in diameter within one decade (by the end of 2006, for funding beginning in [Fiscal Year 1996]). It will also put into place the assets that will expand completeness above 90 percent in the following five years, and extend it both further and to smaller objects in subsequent years. Anticipated cooperation from the Air Force and international programs could shift the attainment of 90 percent completeness forward to 2006, and significantly augment capabilities for orbit determination and physical measurements.

The recommended program requires investment in search telescopes, detectors, and software to fully utilize current technology. The dedicated telescopes of about 2-meter aperture are the core of the search system. One of these is already under construction. State-of-the-art CCD focal plane arrays are required in both telescopes. Acquisition of computers and skilled personnel is required to bring the CCD systems into full operation within three years. One or two existing telescopes near 1-meter aperture, with appropriate advanced focal planes, can round out the survey facilities (capable of both survey work and astrometric follow-up for orbit determination). In addition, enhanced funding is necessary to obtain availability of roughly half time on a 3- to 4-meter class telescope for physical observations of a representative sample of threatening objects. Enhancement of the capability of the Minor Planet Center will be necessary to coordinate the program and handle the enormously enhanced discovery rates.

The level of funding required to carry out the recommended program is as follows:

| Year 1 | Year 2 | Year 3 | Year 4 | Year 5 |
|--------|--------|--------|--------|--------|
| $4.3 M | $4.3 M | $4.7 M | $6.5 M | $4.5 M |

The total cost for 5 years is $24 million. This funding is for the NASA program; funding for the participation of Air Force facilities in NEO surveys is not addressed in this report. Beyond the first 5 years, the annual costs drop down to operations costs of about $3.5 million per year.

It must be emphasized that, without an initial investment in large array CCDs with high quantum efficiency, only moderate improvements will be made over the present rate of discovery of NEO's. Continuation of the present level of NASA supported NEO searches (about $1 million per year) will lead to discovery of about 25 percent of NEOs larger than one kilometer in 10 years and defer 90 percent completeness to some time in the middle of the next century (assuming that the survey would be continued).

# Appendix D

# Close Approach Asteroid Data

The data presented in this appendix is for the more serious student of the asteroid collision problem. Astronomical data is usually presented with orbit semimajor axis and eccentricity as the major descriptors of orbit size and shape. Instead, we have chosen to use period (which is directly related to semimajor axis) and either perihelion or aphelion, depending upon which threatens Earth.

We want to express our appreciation to the astronomers who contributed the data which we modified to produce the following tables.

Table 1, Predicted Close Approaches, was compiled by Don Yeomans of the NASA/Caltech Jet Propulsion Laboratory. He has been involved in interplanetary orbit determination work for a number of years. Tables 2, 3, and 4 were adapted from data provided by David Tholen of the University of Hawaii, Department of Astronomy. Incomparable facilities for ground-based observational astronomy in the optical, infrared, and submillimeter regions of the spectrum reside in Hawaii. The University of Hawaii's facilities are located on Mt. Haleakala on the island of Maui at an elevation of 3,000 meters, and on Mauna Kea on the big island of Hawaii at an elevation of 4,200 meters. The summit of Mauna Kea is internationally recognized as the best observing site in the world. As a consequence, the major telescopes of several nations are located there, and the University of Hawaii is guaranteed access to them.

We want to thank these astronomers for their assistence with this data.

317

## Table D1. Close Encounters with Earth

The following list contains comets and asteroids passing within 0.1 AU of the
Earth through 2020.

Right Ascen. and Declin. are the object's approximate right ascension and declination.

| Object | Date | | Approach Distance | Magnitude (Absolute) | Right. Ascen. | Declin. |
|---|---|---|---|---|---|---|
| 1991 JX | 1995 06 | 9.098 | .0341 | 18.5 | 278 | 37 |
| 2063 Bacchus | 1996 03 | 31.670 | .0678 | 16.4 | 230 | 59 |
| 1991 CS | 1996 08 | 28.419 | .0620 | 17.5 | 53 | − 2 |
| 4197 1982 TA | 1996 10 | 25.639 | .0846 | 14.5 | 289 | 72 |
| 3908 1980 PA | 1996 10 | 27.860 | .0613 | 17.4 | 2 | 32 |
| 1991 VE | 1996 10 | 29.543 | .0853 | 19.0 | 296 | −66 |
| 4179 Toutatis | 1996 11 | 29.953 | .0354 | 15.4 | 204 | −22 |
| 1991 VK | 1997 01 | 10.695 | .0749 | 17.0 | 287 | −18 |
| 6037 1988 EG | 1998 02 | 28.914 | .0318 | 18.7 | 77 | −28 |
| 1991 RB | 1998 09 | 18.475 | .0401 | 19.0 | 170 | −46 |
| 1989 UR | 1998 11 | 28.689 | .0800 | 18.0 | 4 | −18 |
| 1992 SK | 1999 03 | 26.265 | .0560 | 17.5 | 28 | 41 |
| 1991 JX | 1999 06 | 2.819 | .0500 | 18.5 | 291 | 12 |
| 4486 Mithra | 2000 08 | 14.365 | .0466 | 15.4 | 112 | −69 |
| 4179 Toutatis | 2000 10 | 31.186 | .0739 | 15.4 | 218 | −21 |
| 1991 VK | 2002 01 | 16.498 | .0718 | 17.0 | 289 | −24 |
| 4660 Nereus | 2002 01 | 22.512 | .0290 | 18.3 | 287 | −13 |
| 5604 1992 FE | 2002 06 | 22.264 | .0768 | 17.0 | 157 | −47 |
| 1991 BN | 2002 11 | 14.726 | .0775 | 20.0 | 325 | 19 |
| 1990 SM | 2003 02 | 17.275 | .0747 | 16.5 | 59 | 40 |
| 1991 JX | 2003 05 | 20.681 | .0922 | 18.5 | 301 | − 6 |
| 1990 OS | 2003 11 | 11.448 | .0250 | 20.0 | 194 | 40 |
| 1989 QF | 2004 02 | 4.267 | .0748 | 18.0 | 22 | 42 |
| 4179 Toutatis | 2004 09 | 29.567 | .0104 | 15.4 | 218 | −60 |
| 1988 XB | 2004 11 | 21.965 | .0728 | 17.5 | 164 | − 1 |
| 1992 BF | 2005 03 | 3.695 | .0630 | 19.0 | 129 | −62 |
| 1993 VW | 2005 04 | 24.904 | .0862 | 16.5 | 137 | −24 |
| 1992 UY4 | 2005 08 | 8.424 | .0402 | 17.5 | 359 | 12 |
| 1991 RB | 2005 09 | 13.130 | .0785 | 19.0 | 101 | − 1 |
| 1862 Apollo | 2005 11 | 6.802 | .0752 | 16.3 | 154 | 31 |
| Schwassmann- | | | | | | |
| Wachmann 3 | 2006 05 | 10.688 | .0912 | 15.0 | 309 | 21 |
| 1991 VK | 2007 01 | 21.507 | .0679 | 17.0 | 291 | −31 |
| 1862 Apollo | 2007 05 | 8.638 | .0714 | 16.3 | 329 | −39 |
| 2340 Hathor | 2007 10 | 22.238 | .0600 | 20.3 | 278 | −12 |

| Object | Date | | Approach Distance | Magnitude (Absolute) | Right. Ascen. | Declin. |
|---|---|---|---|---|---|---|
| 1989 UR | 2007 11 | 26.357 | .0406 | 18.0 | 117 | −33 |
| 1989 AZ | 2008 01 | 1.478 | .0622 | 19.5 | 185 | 59 |
| 4450 Pan | 2008 02 | 19.932 | .0408 | 17.1 | 74 | −21 |
| 1991 VH | 2008 08 | 12.238 | .0291 | 17.0 | 167 | 9 |
| 4179 Toutatis | 2008 11 | 9.514 | .0502 | 15.4 | 214 | −22 |
| 1993 KH | 2008 11 | 22.328 | .0992 | 19.0 | 331 | −12 |
| 1991 JW | 2009 05 | 23.959 | .0813 | 19.5 | 201 | −23 |
| 1994 CC | 2009 06 | 9.459 | .0163 | 18.0 | 88 | −39 |
| 1991 AQ | 2010 01 | 29.143 | .0892 | 17.5 | 26 | 35 |
| 1989 QF | 2010 08 | 12.405 | .0723 | 18.0 | 49 | − 4 |
| 1991 JW | 2010 11 | 28.916 | .0953 | 19.5 | 326 | − 2 |
| 1990 UN | 2011 03 | 16.824 | .0902 | 23.5 | 94 | 36 |
| 1990 SS | 2011 03 | 17.375 | .0994 | 19.0 | 126 | 50 |
| Honda-Mrkos-Pajdusakova | 2011 08 | 15.275 | .0601 | 18.0 | 92 | −67 |
| 1991 VK | 2012 01 | 25.997 | .0650 | 17.0 | 294 | −37 |
| 4179 Toutatis | 2012 12 | 12.277 | .0463 | 15.4 | 21 | 1 |
| 1988 TA | 2013 05 | 8.082 | .0546 | 21.0 | 304 | 9 |
| 1984 KB | 2013 11 | 11.545 | .0790 | 15.0 | 118 | −52 |
| 2340 Hathor | 2014 10 | 21.895 | .0482 | 20.3 | 89 | 25 |
| 1990 UA | 2015 05 | 13.691 | .0556 | 19.5 | 315 | − 3 |
| 1566 Icarus | 2015 06 | 16.652 | .0538 | 16.4 | 182 | 49 |
| 1994 AW1 | 2015 07 | 15.347 | .0577 | 17.5 | 217 | −23 |
| 1991 VK | 2017 01 | 25.542 | .0647 | 17.0 | 294 | −38 |
| Honda-Mrkos-Pajdusakova | 2017 02 | 11.104 | .0864 | 18.0 | 249 | 23 |
| 5604 1992 FE | 2017 02 | 24.420 | .0336 | 17.0 | 62 | −74 |
| 1984 KB | 2017 05 | 27.630 | .0985 | 15.0 | 183 | 65 |
| 1991 VG | 2017 08 | 7.351 | .0568 | 28.8 | 314 | −43 |
| 3122 Florence | 2017 09 | 1.493 | .0472 | 14.2 | 315 | 2 |
| 5189 1990 UQ | 2017 09 | 26.762 | .0611 | 17.5 | 278 | −71 |
| 1989 UP | 2017 11 | 4.241 | .0471 | 20.5 | 97 | 10 |
| 3361 Orpheus | 2017 11 | 25.690 | .0607 | 19.0 | 340 | −48 |
| 3200 Phaethon | 2017 12 | 16.958 | .0689 | 14.6 | 7 | 27 |
| 1991 VG | 2018 02 | 11.915 | .0473 | 28.8 | 43 | 47 |
| 1981 Midas | 2018 03 | 21.889 | .0896 | 15.0 | 101 | 32 |
| 4581 Asclepius | 2020 03 | 25.616 | .0705 | 20.5 | 84 | 24 |
| 1991 DG | 2020 04 | 6.279 | .0857 | 19.0 | 118 | 51 |
| 1990 MF | 2020 07 | 23.929 | .0546 | 18.7 | 169 | 44 |
| 1988 XB | 2020 11 | 22.708 | .0662 | 17.5 | 163 | − 1 |

# Table D2: ATEN Asteroids

This table includes the orbits of known Atens as of March 21, 1995. The period is given in years, the perihelion is in AU, angles are in degrees, and estimated diameters are in kilometers. The final column gives the spectral type if available. The epoch for all entries is 2449800.5. The authors take responsibilty for any errors in converting the data supplied into the form presented here.

These data are presented in order of decreasing asteroid aphelion. Note: the Earth ranges between 0.984 and 1.16 AU from the sun during the course of a year.

| Designation | Aphelion | period | incl. | Asc. Node | Arg. of Per. | Diam. | Type |
|---|---|---|---|---|---|---|---|
| 3753 1986 TO | 1.511 | 0.997 | 19.81 | 125.7 | 43.62 | 5 | |
| 1994 TF2 | 1.275 | 0.990 | 23.75 | 174.64 | 349.62 | 0.6 | |
| 3362 Khufu 1984 QA | 1.453 | 0.984 | 9.91 | 151.97 | 54.84 | 1 | |
| 5590 1990 VA | 1.260 | 0.978 | 14.19 | 215.72 | 34.35 | 0.4 | |
| 3554 Amun 1986 EB | 1.247 | 0.961 | 23.35 | 358.01 | 359.34 | 3 | |
| 2062 Aten 1976 AA | 1.143 | 0.950 | 18.93 | 107.99 | 147.82 | 1 | S |
| 5381 Sekhmet 1991 JY | 1.228 | 0.922 | 48.97 | 57.88 | 37.41 | 2 | |
| 1993 DA | 1.023 | 0.905 | 12.4 | 328.53 | 353.96 | 0.02 | |
| 5604 1992 FE | 1.303 | 0.893 | 4.78 | 311.43 | 82.36 | 2 | |
| 1989 UQ | 1.158 | 0.876 | 1.28 | 178.03 | 14.62 | 0.6 | |
| 1992 BF | 1.154 | 0.865 | 7.25 | 314.94 | 336.28 | 0.6 | |
| 1995 CR | 1.692 | 0.862 | 4 | 342.33 | 322.1 | 0.2 | |
| 1991 VE | 1.482 | 0.840 | 7.2 | 61.49 | 193.33 | 0.6 | |
| 1993 VD | 1.358 | 0.821 | 2.05 | 2.25 | 253.51 | 0.2 | |
| 2340 Hathor 1976 UA | 1.223 | 0.775 | 5.85 | 210.93 | 39.83 | 0.5 | CSU |
| 2100 Ra-Shalom 1978 RA | 1.195 | 0.759 | 15.75 | 170.24 | 355.95 | 4 | C |
| 1954 XA | 1.046 | 0.686 | 3.91 | 188.77 | 58.73 | 0.6 | |
| 1994 WR12  1.060 | 0.655 | 7.06 | 62.41 | 205.59 | 0.2 | | |
| 1989 VA | 1.162 | 0.622 | 28.79 | 224.95 | 2.79 | 1 | |
| 1994 GL | 1.028 | 0.566 | 3.64 | 196.46 | 179.06 | 0.06 | |
| 1994 XL1 | 1.024 | 0.549 | 28.19 | 252.02 | 356.49 | 0.3 | |

# Table D3: Apollo Asteroids

This table includes the orbits of known Apollos as of March 21, 1995. The period is given in years, the perihelion is in AU, angles are in degrees, and estimated diameters are in kilometers. The final column gives the spectral type if available. The epoch for all entries is 2449800.5. The authors take responsibilty for any errors in converting the data supplied into the form presented here.

These data are presented in order of increasing asteroid perihelion. Note: the Earth ranges between 0.984 and 1.16 AU from the sun during the course of a year.

| Designation | Perigee | Period | Incl. | Asc. Node | Arg. of Per. | Diam. |
|---|---|---|---|---|---|---|
| 3200 Phaethon 1983 TB | 0.1397 | 1.4334 | 22.10 | 264.87 | 321.83 | 5.2 |
| 1566 Icarus 1949 MA | 0.1868 | 1.1193 | 22.88 | 87.47 | 31.20 | 2.0 |
| 5786 Talos 1991 RC | 0.1874 | 1.1246 | 23.24 | 160.67 | 18.26 | 2.0 |
| 1984 QY1 | 0.2195 | 6.8237 | 17.85 | 144.72 | 335.79 | 11.0 |
| 1990 UO | 0.2984 | 1.3710 | 29.34 | 205.06 | 332.95 | 0.3 |
| 2212 Hephaistos 1978 SB | 0.3610 | 3.1924 | 11.77 | 27.75 | 208.36 | 5.0 |
| 1991 LH | 0.3644 | 1.5720 | 52.06 | 280.16 | 203.92 | 1.5 |
| 1994 PM | 0.3660 | 1.7959 | 17.94 | 139.38 | 303.27 | 1.0 |
| 1991 TB2 | 0.3951 | 3.7127 | 8.62 | 296.41 | 195.66 | 1.5 |
| 5143 Heracles 1991 VL | 0.4194 | 2.4833 | 9.18 | 310.04 | 226.39 | 6.0 |
| 5660 1974 MA | 0.4237 | 2.3870 | 37.94 | 301.85 | 126.71 | 3.0 |
| 1995 CS | 0.4359 | 2.7021 | 2.61 | 135.06 | 252.05 | 0.0 |
| 2101 Adonis 1936 CA | 0.4407 | 2.5650 | 1.34 | 350.02 | 42.30 | 0.7 |
| 5025 P-L | 0.4432 | 8.6348 | 6.34 | 354.67 | 151.44 | 5.0 |
| 3838 Epona 1986 WA | 0.4487 | 1.8458 | 29.28 | 235.06 | 49.43 | 3.0 |
| 1993 HD | 0.4822 | 1.7140 | 5.74 | 201.77 | 252.55 | 0.1 |
| 1990 SM | 0.4844 | 3.1673 | 11.56 | 137.28 | 105.91 | 2.0 |
| 1991 AQ | 0.4951 | 3.3100 | 3.21 | 342.06 | 239.69 | 1.5 |
| 1993 KA2 | 0.5023 | 3.3241 | 3.18 | 238.78 | 261.44 | 0.0 |
| 1995 EK1 | 0.5074 | 3.3770 | 8.81 | 355.03 | 296.53 | 1.0 |
| 5828 1991 AM | 0.5166 | 2.2120 | 30.04 | 124.91 | 152.53 | 2.0 |
| 4197 1982 TA | 0.5222 | 3.4818 | 12.20 | 9.50 | 119.23 | 5.0 |
| 6063 1984 KB | 0.5222 | 3.2981 | 4.85 | 169.25 | 336.49 | 4.0 |
| 5693 1993 EA | 0.5275 | 1.4348 | 5.05 | 96.60 | 258.56 | 1.5 |
| 4769 Castalia 1989 PB | 0.5494 | 1.0961 | 8.88 | 325.04 | 121.22 | 1.6 |
| 4953 1990 MU | 0.5561 | 2.0635 | 24.42 | 77.42 | 77.38 | 5.0 |
| 1993 PB | 0.5602 | 1.6978 | 40.84 | 315.32 | 212.21 | 2.0 |
| 1864 Daedalus 1971 FA | 0.5630 | 1.7658 | 22.17 | 6.09 | 325.43 | 3.0 |
| 1991 WA | 0.5631 | 1.9770 | 39.65 | 66.07 | 241.73 | 2.0 |
| 1865 Cerberus 1971 UA | 0.5758 | 1.1225 | 16.09 | 212.35 | 325.13 | 1.0 |
| 1994 ES1 | 0.5792 | 1.6773 | 0.94 | 352.34 | 279.11 | 0.0 |
| 1995 DW1 | 0.5848 | 1.0614 | 15.10 | 348.26 | 326.88 | 0.2 |

| Designation | Perigee | Period | Incl. | Asc. Node | Arg. of Per. | Diam. |
|---|---|---|---|---|---|---|
| 4034 1986 PA | 0.5892 | 1.0912 | 11.17 | 157.44 | 296.44 | 1.0 |
| 4341 Poseidon 1987 KF | 0.5898 | 2.4872 | 11.86 | 107.55 | 15.47 | 3.0 |
| 4450 Pan 1987 SY | 0.5962 | 1.7312 | 5.51 | 311.52 | 291.38 | 1.5 |
| 1987 OA | 0.6051 | 1.8301 | 9.02 | 179.69 | 235.38 | 0.8 |
| 1993 PC | 0.6068 | 1.2402 | 4.15 | 336.85 | 168.12 | 1.0 |
| 1992 HF | 0.6095 | 1.6400 | 13.30 | 212.91 | 128.05 | 0.3 |
| Hermes 1937 UB | 0.6197 | 2.1123 | 6.13 | 34.11 | 91.85 | 1.0 |
| 1981 Midas 1973 EA | 0.6216 | 2.3670 | 39.83 | 356.48 | 267.68 | 4.0 |
| 1995 BL2 | 0.6228 | 1.3666 | 23.08 | 311.72 | 348.22 | 1.0 |
| 2201 Oljato 1947 XC | 0.6297 | 3.2097 | 2.51 | 76.34 | 95.81 | 2.0 |
| 3360 1981 VA | 0.6334 | 3.8690 | 21.74 | 244.81 | 60.58 | 2.0 |
| 6037 1988 EG | 0.6353 | 1.4295 | 3.48 | 182.27 | 241.46 | 0.7 |
| 1994 XD | 0.6391 | 3.6232 | 4.34 | 96.96 | 247.73 | 0.6 |
| 5131 1990 BG | 0.6392 | 1.8126 | 36.38 | 109.83 | 135.69 | 6.0 |
| 1979 XB | 0.6468 | 3.4025 | 24.85 | 85.14 | 75.64 | 0.6 |
| 1862 Apollo 1932 HA | 0.6476 | 1.7843 | 6.35 | 35.27 | 285.58 | 1.5 |
| 4581 Asclepius 1989 FC | 0.6574 | 1.0341 | 4.91 | 179.85 | 255.00 | 0.3 |
| 1991 GO | 0.6627 | 2.7427 | 9.66 | 24.30 | 88.58 | 0.6 |
| 6239 1989 QF | 0.6762 | 1.2353 | 3.93 | 344.15 | 239.50 | 0.9 |
| 1991 CB1 | 0.6839 | 2.1912 | 14.56 | 316.85 | 345.53 | 1.0 |
| 1989 UR | 0.6953 | 1.1225 | 10.34 | 233.89 | 289.29 | 1.0 |
| 2063 Bacchus 1977 HB | 0.7012 | 1.1191 | 9.41 | 32.63 | 55.05 | 2.0 |
| 1991 BA | 0.7072 | 3.1860 | 2.12 | 118.25 | 71.86 | 0.0 |
| 4183 Cuno 1959 LM | 0.7193 | 2.7880 | 6.76 | 295.17 | 235.22 | 5.0 |
| 1994 CK1 | 0.7211 | 2.6142 | 4.40 | 328.11 | 27.67 | 1.5 |
| 1992 TB | 0.7216 | 1.5543 | 28.31 | 185.02 | 5.92 | 1.0 |
| 1994 AH2 | 0.7289 | 4.0135 | 9.63 | 163.68 | 24.82 | 2.0 |
| 1993 WD | 0.7382 | 1.0099 | 63.46 | 55.89 | 132.30 | 2.0 |
| 4486 Mithra 1987 SB | 0.7427 | 3.2658 | 3.04 | 81.94 | 168.47 | 3.0 |
| 1992 LC | 0.7434 | 3.9987 | 17.84 | 61.28 | 89.64 | 4.0 |
| 1991 RB | 0.7484 | 1.7464 | 19.53 | 358.88 | 68.71 | 0.6 |
| 1988 XB | 0.7607 | 1.7776 | 3.12 | 73.02 | 279.79 | 1.0 |
| 1992 QN | 0.7629 | 1.2994 | 9.58 | 355.40 | 202.11 | 2.0 |
| 1990 TG1 | 0.7631 | 3.9171 | 9.06 | 204.37 | 33.28 | 4.0 |
| 1983 LC | 0.7657 | 4.2661 | 1.52 | 159.07 | 184.67 | 0.6 |
| 1685 Toro 1948 OA | 0.7710 | 1.5983 | 9.37 | 273.70 | 126.95 | 12.0 |
| 1990 UA | 0.7713 | 2.2581 | 0.96 | 102.09 | 203.87 | 0.5 |
| 1994 GK | 0.7718 | 2.7459 | 5.66 | 14.67 | 111.58 | 0.1 |
| 4544 Xanthus 1989 FB | 0.7813 | 1.0637 | 14.14 | 23.44 | 333.62 | 1.5 |
| 5731 1988 VP4 | 0.7862 | 3.4047 | 11.64 | 282.06 | 215.63 | 3.0 |
| 1994 EK | 0.7863 | 2.8267 | 5.76 | 333.54 | 98.01 | 0.4 |
| 1990 HA | 0.7908 | 4.1386 | 3.88 | 184.43 | 307.89 | 1.5 |
| 2135 Aristaeus 1977 HA | 0.7949 | 2.0233 | 23.04 | 190.74 | 290.63 | 1.0 |
| 1994 XG | 0.8017 | 1.9708 | 11.32 | 231.10 | 46.14 | 1.0 |
| 1983 VA | 0.8025 | 4.2180 | 16.24 | 76.81 | 11.67 | 2.0 |
| 1988 TA | 0.8034 | 1.9126 | 2.54 | 194.49 | 104.57 | 0.2 |

| Designation | Perigee | Period | Incl. | Asc. Node | Arg. of Per. | Diam. |
|---|---|---|---|---|---|---|
| 1994 NE | 0.8052 | 2.9051 | 27.53 | 104.29 | 246.20 | 0.4 |
| 1990 UN | 0.8073 | 2.2347 | 3.66 | 7.62 | 97.08 | 0.1 |
| 5189 1990 UQ | 0.8098 | 1.9312 | 3.58 | 134.80 | 159.37 | 1.0 |
| 5011 Ptah 6743 P-L | 0.8177 | 2.0906 | 7.39 | 10.38 | 105.36 | 1.5 |
| 3361 Orpheus 1982 HR | 0.8192 | 1.3295 | 2.68 | 189.11 | 301.53 | 0.6 |
| 2329 Orthos 1976 WA | 0.8199 | 3.7236 | 24.41 | 168.84 | 145.75 | 4.0 |
| 1993 UC | 0.8226 | 3.8067 | 25.99 | 165.41 | 322.95 | 3.0 |
| 1994 LX | 0.8245 | 1.4167 | 36.90 | 110.65 | 349.01 | 4.0 |
| 1993 VA | 0.8255 | 1.5789 | 7.26 | 132.60 | 336.35 | 1.5 |
| 1620 Geographos 1951 RA | 0.8278 | 1.3900 | 13.33 | 336.65 | 276.74 | 2.3 |
| 5645 1990 SP | 0.8300 | 1.5769 | 13.51 | 45.23 | 47.96 | 2.0 |
| 1950 DA | 0.8381 | 2.1853 | 12.08 | 356.28 | 224.26 | 3.0 |
| 1993 GD | 0.8399 | 1.1572 | 15.45 | 200.87 | 201.92 | 0.3 |
| 1992 SK | 0.8429 | 1.3950 | 15.31 | 8.36 | 233.48 | 1.0 |
| 1991 EE | 0.8436 | 3.3660 | 9.76 | 168.46 | 115.03 | 1.0 |
| 1993 KH | 0.8501 | 1.3705 | 12.80 | 53.86 | 293.57 | 0.6 |
| 1991 BB | 0.8630 | 1.2921 | 38.47 | 294.34 | 322.81 | 2.0 |
| 1991 BN | 0.8686 | 1.7327 | 3.44 | 268.46 | 80.54 | 0.4 |
| 1992 CC1 | 0.8698 | 1.6413 | 36.88 | 348.61 | 21.90 | 4.0 |
| 1866 Sisyphus 1972 XA | 0.8726 | 2.6051 | 41.16 | 63.01 | 292.97 | 9.6 |
| 1993 VW | 0.8741 | 2.2066 | 8.68 | 230.58 | 280.93 | 2.0 |
| 1989 AZ | 0.8752 | 2.1112 | 11.76 | 295.06 | 111.63 | 0.5 |
| 4257 Ubasti 1987 QA | 0.8756 | 2.1139 | 40.70 | 168.65 | 278.89 | 2.0 |
| 5496 1973 NA | 0.8805 | 3.7967 | 68.02 | 100.38 | 118.24 | 4.0 |
| 1978 CA | 0.8830 | 1.1926 | 26.11 | 160.63 | 102.12 | 1.0 |
| 1993 TZ | 0.8837 | 2.8784 | 4.16 | 202.96 | 231.23 | 0.0 |
| 1863 Antinous 1948 EA | 0.8900 | 3.3975 | 18.41 | 346.84 | 266.91 | 2.0 |
| 1986 JK | 0.8930 | 4.6753 | 2.13 | 62.17 | 232.40 | 0.6 |
| 1990 SS | 0.8944 | 2.2222 | 19.39 | 359.43 | 115.74 | 0.6 |
| 1994 RB | 0.8973 | 3.9251 | 26.62 | 338.94 | 52.47 | 0.1 |
| 1994 XM1 | 0.9005 | 3.0180 | 4.11 | 76.35 | 40.66 | 0.0 |
| 1990 OS | 0.9030 | 2.1564 | 1.10 | 347.40 | 20.03 | 0.4 |
| 1994 RC | 0.9030 | 3.4151 | 4.72 | 345.45 | 284.34 | 0.6 |
| 1994 PC1 | 0.9047 | 1.5611 | 33.50 | 117.29 | 47.53 | 2.0 |
| 2102 Tantalus 1975 YA | 0.9050 | 1.4653 | 64.01 | 93.70 | 61.60 | 3.3 |
| 3103 Eger 1982 BB | 0.9074 | 1.6672 | 20.93 | 129.22 | 253.79 | 3.0 |
| 1991 DG | 0.9092 | 1.7048 | 11.15 | 179.63 | 63.05 | 0.6 |
| 1991 VK | 0.9109 | 2.5032 | 5.41 | 294.37 | 173.25 | 1.5 |
| 1994 VH8 | 0.9113 | 2.0916 | 3.37 | 37.78 | 314.14 | 0.0 |
| 1989 JA | 0.9133 | 2.3556 | 15.23 | 60.93 | 231.80 | 1.5 |
| 4179 Toutatis 1989 AC | 0.9154 | 3.9894 | 0.47 | 128.30 | 274.08 | 3.3 |
| 1991 JW | 0.9155 | 1.0580 | 8.71 | 53.41 | 301.76 | 0.5 |
| 1993 VB | 0.9173 | 2.6393 | 5.06 | 145.23 | 322.63 | 0.5 |
| 1992 BC | 0.9210 | 1.6805 | 14.21 | 122.82 | 77.07 | 0.6 |
| 1994 UG | 0.9246 | 1.3572 | 4.49 | 11.73 | 225.91 | 0.2 |
| 1991 VA | 0.9264 | 1.7079 | 6.52 | 36.97 | 313.41 | 0.0 |

| Designation | Perigee | Period | Incl. | Asc. Node | Arg. of Per. | Diam. |
|---|---|---|---|---|---|---|
| 1993 BW2 | 0.9265 | 1.5428 | 21.91 | 120.50 | 287.37 | 1.0 |
| 1991 CS | 0.9381 | 1.1899 | 37.10 | 156.23 | 249.26 | 1.0 |
| 1991 TU | 0.9417 | 1.6686 | 7.55 | 192.73 | 222.09 | 0.0 |
| 6047 1991 TB1 | 0.9419 | 1.7527 | 23.46 | 5.57 | 103.55 | 1.5 |
| 1990 MF | 0.9510 | 2.3087 | 1.86 | 209.87 | 113.89 | 0.8 |
| 1994 CN2 | 0.9519 | 1.9732 | 1.43 | 98.93 | 247.89 | 2.0 |
| 4660 Nereus 1982 DB | 0.9531 | 1.8182 | 1.41 | 313.97 | 157.96 | 0.8 |
| 1994 CC | 0.9545 | 2.0945 | 4.63 | 268.11 | 24.63 | 1.0 |
| 1983 VB | 0.9546 | 2.4531 | 12.05 | 247.70 | 115.36 | 0.8 |
| 1992 DU | 0.9571 | 1.2490 | 25.05 | 337.28 | 121.63 | 0.0 |
| 1992 HE | 0.9584 | 3.3530 | 37.36 | 26.62 | 262.57 | 6.0 |
| 1991 TF3 | 0.9588 | 2.9167 | 14.04 | 6.02 | 303.20 | 0.6 |
| 1993 UA | 0.9608 | 2.8712 | 4.58 | 26.48 | 330.11 | 0.0 |
| 1995 DV1 | 0.9643 | 5.5261 | 3.69 | 171.63 | 283.97 | 0.1 |
| 1994 GV | 0.9687 | 2.8923 | 0.45 | 19.31 | 154.30 | 0.0 |
| 6344 P-L | 0.9696 | 4.2578 | 4.47 | 180.50 | 237.09 | 0.2 |
| 1991 VH | 0.9730 | 1.2113 | 13.91 | 138.82 | 206.99 | 1.5 |
| 1991 XA | 0.9748 | 3.4178 | 5.28 | 76.17 | 309.11 | 0.1 |
| 1993 HP1 | 0.9750 | 2.8104 | 8.00 | 36.37 | 152.25 | 0.0 |
| 1991 VG | 0.9765 | 1.0406 | 1.44 | 73.51 | 24.15 | 0.0 |
| 1993 HC | 0.9807 | 2.8047 | 9.39 | 200.85 | 306.32 | 0.3 |
| 1989 UP | 0.9818 | 2.5443 | 3.86 | 52.79 | 17.21 | 0.3 |
| 1994 CB | 0.9825 | 1.2318 | 18.25 | 310.04 | 288.40 | 0.2 |
| 1993 XN2 | 0.9833 | 3.0809 | 25.38 | 59.06 | 312.89 | 2.0 |
| 3752 Camillo 1985 PA | 0.9863 | 1.6807 | 55.54 | 147.33 | 312.21 | 3.0 |
| 1989 DA | 0.9875 | 3.1796 | 6.44 | 349.08 | 138.73 | 1.0 |
| 1992 SY | 0.9933 | 3.2825 | 8.02 | 5.62 | 114.98 | 1.0 |
| 1994 EU | 0.9952 | 1.6079 | 6.44 | 350.96 | 145.77 | 0.0 |
| 1992 JB | 0.9964 | 1.9417 | 16.07 | 217.81 | 306.75 | 1.0 |
| 4015 Wilson-Harrington 1979 VA | 0.9989 | 4.2939 | 2.78 | 270.14 | 91.07 | 4.0 |
| 1991 TT | 1.0014 | 1.3029 | 14.75 | 191.77 | 218.11 | 0.0 |
| 1992 JD | 1.0020 | 1.0525 | 13.54 | 221.92 | 285.87 | 0.0 |
| 1993 BX3 | 1.0034 | 1.6482 | 2.79 | 174.99 | 289.77 | 0.3 |
| 3671 Dionysus 1984 KD | 1.0034 | 3.2525 | 13.62 | 81.76 | 203.63 | 2.0 |
| 1994 CJ1 | 1.0051 | 1.8202 | 2.31 | 171.61 | 64.92 | 0.2 |
| 1989 VB | 1.0055 | 2.5465 | 2.13 | 38.34 | 329.60 | 0.4 |
| 1992 YD3 | 1.0061 | 1.2592 | 27.04 | 273.63 | 173.75 | 0.0 |
| 1993 KA | 1.0074 | 1.4063 | 6.05 | 235.17 | 341.90 | 0.0 |
| 1992 UY4 | 1.0092 | 4.3244 | 2.83 | 308.41 | 37.43 | 1.0 |
| 1993 QA | 1.0099 | 1.7945 | 12.63 | 146.09 | 323.25 | 1.0 |
| 1991 JX | 1.0101 | 3.9980 | 2.31 | 212.16 | 64.47 | 0.8 |
| 6053 1993 BW3 | 1.0119 | 3.1468 | 21.59 | 317.96 | 74.57 | 5.0 |
| 1994 FA | 1.0142 | 2.2869 | 13.04 | 355.09 | 154.66 | 0.0 |
| 1993 FA1 | 1.0146 | 1.7032 | 20.46 | 186.67 | 343.60 | 0.0 |
| 3757 1982 XB | 1.0157 | 2.4857 | 3.87 | 74.50 | 16.77 | 0.5 |

# Table D4: Amor Asteroids

This table includes the orbits of known Amors as of March 21, 1995. The period is given in years, the perihelion is in AU, angles are in degrees, and estimated diameters are in kilometers. The final column gives the spectral type if available. The epoch for all entries is 2449800.5. The authors take responsibilty for any errors in converting the data supplied into the form presented here.

These data are presented in order of increasing asteroid perihelion. Note: the Earth ranges between 0.984 and 1.16 AU from the sun during the course of a year.

| Designation | Period | Perihelion | incl. | Arg. of Per. | Asc. Node | Diam. | Type |
|---|---|---|---|---|---|---|---|
| 3122 Florence 1981 ET3 | 2.352 | 1.0210 | 22.17 | 335.53 | 27.55 | 6.0 | |
| 1994 AW1 | 1.161 | 1.0215 | 24.09 | 289.74 | 37.13 | 2.0 | |
| 1993 BD3 | 2.090 | 1.0220 | 0.88 | 312.96 | 168.86 | 0.0 | |
| 1991 DB | 2.248 | 1.0265 | 11.43 | 157.78 | 50.98 | 0.8 | |
| 1993 OM7 | 1.551 | 1.0275 | 25.95 | 296.98 | 142.47 | 0.8 | |
| 1991 OA | 3.972 | 1.0343 | 5.51 | 305.88 | 317.27 | 1.0 | |
| 1991 FB | 3.643 | 1.0351 | 9.19 | 18.41 | 218.33 | 0.6 | |
| 1993 DQ1 | 2.913 | 1.0381 | 9.97 | 313.03 | 344.41 | 2.0 | |
| 1991 JR | 1.662 | 1.0386 | 10.11 | 59.50 | 207.11 | 0.1 | |
| 3908 1980 PA | 2.669 | 1.0415 | 2.17 | 261.18 | 125.72 | 0.8 | V |
| 2608 Seneca 1978DA | 3.931 | 1.0421 | 15.34 | 168.91 | 33.90 | 1.0 | S |
| 1987 WC | 1.589 | 1.0438 | 15.83 | 51.27 | 308.10 | 0.5 | |
| 1994 ND | 3.188 | 1.0485 | 27.19 | 102.14 | 227.88 | 1.0 | |
| 2061 Anza 1960 UA | 3.410 | 1.0500 | 3.76 | 207.07 | 156.42 | 3.0 | TCG |
| 1992 NA | 3.697 | 1.0513 | 9.75 | 348.91 | 7.85 | 2.0 | |
| 5797 1980 AA | 2.603 | 1.0520 | 4.18 | 298.41 | 168.13 | 0.6 | |
| 1987 SF3 | 3.384 | 1.0520 | 3.32 | 187.02 | 133.63 | 0.8 | |
| 3988 1986 LA | 1.920 | 1.0556 | 10.77 | 229.30 | 86.62 | 0.8 | |
| 1993 UD | 1.516 | 1.0632 | 22.78 | 24.46 | 254.66 | 0.4 | |
| 1943 Anteros 1973 EC | 1.710 | 1.0642 | 8.70 | 245.73 | 338.20 | 2.0 | S |
| 1917 Cuyo 1968 AA | 3.155 | 1.0673 | 23.94 | 187.77 | 194.26 | 6.0 | |
| 1994 PC | 1.964 | 1.0707 | 9.45 | 123.89 | 256.52 | 2.0 | |
| 3551 Verenia 1983 RD | 3.028 | 1.0746 | 9.50 | 173.29 | 193.06 | 1.0 | V |
| 4596 1981 QB | 3.351 | 1.0775 | 37.12 | 153.80 | 248.30 | 2.0 | |
| 1915 Quetzalcoatl 1953 EA | 4.039 | 1.0802 | 20.46 | 162.36 | 347.90 | 0.5 | SMU |
| 5587 1990 SB | 3.700 | 1.0806 | 18.09 | 189.88 | 86.19 | 7.0 | |
| 1992 LR | 2.478 | 1.0821 | 2.02 | 232.38 | 67.85 | 1.0 | |
| 1221 Amor 1932 EA1 | 2.660 | 1.0840 | 11.89 | 170.83 | 26.30 | 1.0 | |
| 1994 NK | 3.606 | 1.0847 | 5.67 | 119.54 | 128.48 | 0.4 | |
| 4688 1980 WF | 3.341 | 1.0854 | 6.41 | 241.03 | 212.91 | 0.5 | QU |
| 1980 Tezcatlipoca 1950 LA | 2.235 | 1.0858 | 26.85 | 246.03 | 115.29 | 13.0 | SU |

| Designation | Period | Perihelion | incl. | Arg. of Per. | Asc. Node | Diam. | Type |
|---|---|---|---|---|---|---|---|
| 1988 SM | 2.144 | 1.0920 | 10.92 | 0.38 | 312.92 | 1.0 | |
| 1995 BC2 | 2.655 | 1.0922 | 5.02 | 328.01 | 81.19 | 2.0 | |
| 1994 PN | 3.662 | 1.0928 | 46.05 | 112.54 | 233.88 | 2.0 | |
| 1992 SL | 2.103 | 1.0934 | 8.59 | 0.41 | 344.50 | 1.0 | |
| 1993 HA | 1.445 | 1.0939 | 7.73 | 182.76 | 263.47 | | |
| 1991 FA | 2.784 | 1.0950 | 3.07 | 338.84 | 91.76 | 2.0 | |
| 1994 TW1 | 4.171 | 1.0966 | 36.03 | 2.92 | 62.19 | 5.0 | |
| 5863 1983 RB | 3.312 | 1.0972 | 19.43 | 168.79 | 114.78 | 3.0 | |
| 887 Alinda | 3.936 | 1.0983 | 9.27 | 110.09 | 349.81 | 5.4 | S |
| 1995 BK2 | 3.815 | 1.0985 | 24.73 | 130.51 | 349.56 | 0.1 | |
| 1989 ML | 1.435 | 1.0988 | 4.37 | 103.83 | 183.10 | 0.5 | |
| 1990 UP | 1.526 | 1.1020 | 28.06 | 32.59 | 293.85 | 0.3 | |
| 3288 Seleucus 1982 DV | 2.898 | 1.1030 | 5.93 | 218.11 | 349.24 | 3.0 | S |
| 4954 Eric 1990 SQ | 2.833 | 1.1049 | 17.47 | 358.12 | 52.01 | 12.0 | |
| 1994 EF2 | 3.470 | 1.1056 | 23.31 | 345.75 | 123.68 | 2.0 | |
| 1972 RB | 3.152 | 1.1059 | 5.22 | 176.81 | 152.34 | 0.6 | |
| 1992 TC | 1.959 | 1.1080 | 7.08 | 88.11 | 275.36 | 1.0 | |
| 1982 YA | 7.103 | 1.1111 | 34.90 | 268.91 | 143.80 | 4.0 | |
| 4401 Aditi 1985 TB | 4.135 | 1.1139 | 26.79 | 23.31 | 67.08 | 3.0 | |
| 2202 Pele 1972 RA | 3.465 | 1.1164 | 8.78 | 169.71 | 217.20 | 2.0 | |
| 1990 SA | 2.740 | 1.1170 | 37.53 | 171.69 | 114.31 | 2.0 | |
| 1985 WA | 4.772 | 1.1174 | 9.78 | 43.00 | 351.11 | 0.8 | |
| 3553 Mera 1985 JA | 2.109 | 1.1178 | 36.76 | 231.95 | 288.85 | 2.0 | |
| 1580 Betulia 1950 KA | 3.251 | 1.1190 | 52.12 | 61.68 | 159.31 | 7.6 | C |
| 1991 JG1 | 1.609 | 1.1200 | 33.85 | 225.78 | 322.57 | 0.6 | |
| 1993 RA | 2.676 | 1.1207 | 5.70 | 171.27 | 265.28 | 0.8 | |
| 1987 QB | 4.667 | 1.1241 | 3.48 | 152.88 | 156.10 | 0.6 | |
| 1627 Ivar 1929 SH | 2.543 | 1.1241 | 8.44 | 132.61 | 167.41 | 7.0 | S |
| 1989 RC | 3.516 | 1.1242 | 7.38 | 139.68 | 181.09 | 1.0 | |
| 3199 Nefertiti 1982 RA | 1.975 | 1.1277 | 32.96 | 339.41 | 53.32 | 3.0 | S |
| 1977 VA | 2.545 | 1.1297 | 2.98 | 223.94 | 172.37 | 0.6 | |
| 433 Eros | 1.761 | 1.1332 | 10.82 | 303.71 | 178.60 | 2.0 | S |
| 5324 Lyapunov 1987 SL | 5.090 | 1.1387 | 19.48 | 352.42 | 320.15 | 6.0 | |
| 4947 Ninkasi 1988 TJ1 | 1.603 | 1.1392 | 15.65 | 214.84 | 192.74 | 0.8 | |
| 5836 1993 MF | 3.820 | 1.1439 | 8.03 | 240.43 | 74.78 | 6.0 | |
| 1994 JX | 4.529 | 1.1515 | 32.68 | 52.23 | 192.21 | 1.0 | |
| 1993 TQ2 | 2.799 | 1.1525 | 6.04 | 13.01 | 77.24 | 0.4 | |
| 1990 BA | 2.296 | 1.1528 | 1.99 | 311.17 | 170.79 | 1.0 | |
| 1990 VB | 3.819 | 1.1540 | 14.56 | 253.91 | 102.25 | 2.0 | |
| 5879 1992 CH1 | 2.071 | 1.1545 | 21.57 | 145.27 | 355.45 | 1.0 | |
| 1993 HO1 | 2.801 | 1.1595 | 5.90 | 22.23 | 104.98 | 2.0 | |
| 1994 BB | 2.876 | 1.1613 | 1.14 | 122.19 | 335.84 | 0.1 | |
| 1994 AB1 | 4.794 | 1.1614 | 4.52 | 66.49 | 342.39 | 2.0 | |
| 1994 QC | 1.524 | 1.1682 | 13.87 | 161.92 | 94.08 | 0.6 | |
| 1993 BU3 | 3.732 | 1.1689 | 5.29 | 315.61 | 144.39 | 0.2 | |

| Designation | Period | Perihelion | incl. | Arg. of Per. | Asc. Node | Diam. | Type |
|---|---|---|---|---|---|---|---|
| 1986 NA | 3.100 | 1.1704 | 10.34 | 243.21 | 35.68 | 0.4 | |
| 1992 JE | 3.240 | 1.1753 | 5.86 | 193.25 | 109.45 | 2.0 | |
| 1992 SZ | 3.212 | 1.1757 | 9.27 | 3.74 | 314.60 | 0.4 | |
| 5332 1990 DA | 3.182 | 1.1761 | 25.43 | 142.48 | 305.56 | 7.0 | |
| 1994 US | 4.481 | 1.1767 | 8.46 | 223.06 | 121.45 | 0.2 | |
| 6178 1986 DA | 4.737 | 1.1781 | 4.29 | 64.41 | 126.87 | 4.0 | |
| 4788 P-L | 4.259 | 1.1821 | 10.98 | 176.92 | 97.14 | 2.0 | |
| 3352 McAuliffe 1981 CW | 2.574 | 1.1845 | 4.77 | 106.87 | 15.59 | 3.0 | |
| 1987 PA | 4.479 | 1.1847 | 16.35 | 307.97 | 337.77 | 0.8 | |
| 1989 OB | 4.427 | 1.1880 | 7.91 | 289.02 | 71.72 | 2.0 | |
| 1977 QQ | 5.320 | 1.1880 | 25.20 | 133.83 | 247.77 | 4.0 | |
| 3102 Krok 1981 QA | 3.158 | 1.1891 | 8.41 | 171.70 | 154.31 | 1.0 | QRS |
| 719 Albert | 4.154 | 1.1904 | 11.24 | 183.96 | 155.07 | 2.0 | |
| 1989 RS1 | 3.497 | 1.1941 | 7.18 | 174.03 | 180.86 | 1.0 | |
| 1983 LB | 3.457 | 1.1958 | 25.35 | 80.82 | 220.18 | 2.0 | |
| 5626 1991 FE | 3.253 | 1.1981 | 3.86 | 172.87 | 231.07 | 4.0 | |
| 1994 LW | 5.624 | 1.2033 | 23.02 | 240.44 | 54.41 | 2.0 | |
| 5646 1990 TR | 3.135 | 1.2049 | 7.90 | 13.61 | 335.37 | 5.0 | |
| 1992 AA | 2.789 | 1.2094 | 8.29 | 102.15 | 354.30 | 2.0 | |
| 3552 Don Quixote 1983 SA | 8.712 | 1.2105 | 30.78 | 350.02 | 316.59 | 18.0 | D |
| 1988 NE | 3.221 | 1.2160 | 9.93 | 253.53 | 354.81 | 0.8 | |
| 4487 Pocahontas 1987 UA | 2.276 | 1.2170 | 16.40 | 197.55 | 173.72 | 1.0 | |
| 5751 1992 AC | 3.053 | 1.2183 | 16.05 | 121.12 | 25.13 | 8.0 | |
| 1991 XB | 5.080 | 1.2214 | 16.29 | 249.74 | 172.00 | 1.0 | |
| 4957 Brucemurray 1990 XJ | 1.959 | 1.2227 | 35.01 | 254.29 | 97.46 | 4.0 | |
| 1993 QP | 3.505 | 1.2244 | 7.24 | 296.70 | 46.55 | 1.0 | |
| 4055 Magellan 1985 DO2 | 2.456 | 1.2262 | 23.24 | 164.29 | 154.11 | 3.0 | V |
| 1036 Ganymede | 4.338 | 1.2290 | 26.63 | 215.11 | 132.27 | 41.0 | S |
| 1993 UB | 3.437 | 1.2293 | 25.02 | 30.83 | 20.78 | 2.0 | |
| 5869 Tanith 1988 VN4 | 2.439 | 1.2304 | 17.94 | 227.37 | 230.55 | 2.0 | |
| 5370 Taranis 1986R | 6.123 | 1.2323 | 19.01 | 177.19 | 161.11 | 5.0 | |
| 2368 Beltrovata 1977 RA | 3.053 | 1.2337 | 5.24 | 287.05 | 42.24 | 3.0 | SQ |
| 6050 1992 AE | 3.269 | 1.2417 | 6.40 | 87.90 | 284.53 | 3.0 | |
| 1990 KA | 3.260 | 1.2471 | 7.56 | 105.10 | 146.50 | 2.0 | |
| 5620 1990 OA | 3.173 | 1.2475 | 7.84 | 128.31 | 152.95 | 1.5 | |
| 5653 1992 WD5 | 2.403 | 1.2491 | 6.86 | 9.50 | 122.15 | 3.0 | |
| 1994 GY | 4.363 | 1.2500 | 12.46 | 33.43 | 189.95 | 2.0 | |
| 1992 BA | 1.553 | 1.2505 | 10.48 | 139.64 | 107.17 | 0.3 | |
| 1916 Boreas 1953 RA | 3.427 | 1.2516 | 12.83 | 340.20 | 335.27 | 3.0 | S |
| 2059 Baboquivari 1963 UA | 4.311 | 1.2546 | 11.00 | 200.43 | 191.17 | 2.0 | |
| 1994 RH | 3.367 | 1.2550 | 18.92 | 330.99 | 91.81 | 3.0 | |
| 1994 TA2 | 4.383 | 1.2555 | 7.07 | 200.47 | 119.12 | 0.3 | |
| 1994 TE2 | 3.498 | 1.2563 | 5.70 | 198.07 | 182.14 | 0.2 | |
| 1991 NT3 | 2.437 | 1.2602 | 13.86 | 286.78 | 292.75 | 5.0 | |
| 1991 RJ2 | 3.287 | 1.2646 | 8.91 | 171.39 | 150.33 | 0.6 | |

| Designation | Period | Perihelion | incl. | Arg. of Per. | Asc. Node | Diam. | Type |
|---|---|---|---|---|---|---|---|
| 1993 MO | 2.074 | 1.2671 | 22.63 | 110.90 | 167.06 | 2.0 | |
| 3691 1982 FT | 2.363 | 1.2706 | 20.37 | 348.24 | 234.61 | 5.0 | |
| 3271 1982 RB | 3.048 | 1.2722 | 25.00 | 158.35 | 158.66 | 2.0 | |
| 1988 PA | 3.153 | 1.2736 | 8.21 | 161.74 | 136.95 | 2.0 | |
| 1992 UB | 5.368 | 1.2806 | 15.94 | 73.68 | 290.67 | 2.0 | |
| 1991 PM5 | 2.255 | 1.2808 | 14.42 | 132.09 | 140.27 | 1.0 | |
| 1993 FS | 3.323 | 1.2810 | 10.13 | 178.73 | 20.82 | 0.4 | |
| 1993 BD2 | 3.093 | 1.2880 | 25.59 | 96.50 | 65.06 | 0.6 | |
| 1992 BL2 | 2.180 | 1.2947 | 36.86 | 297.15 | 23.40 | 4.0 | |
| 1993 VC | 4.621 | 1.2956 | 3.20 | 241.94 | 177.04 | 0.3 | |
| 1992 OM | 3.249 | 1.2970 | 8.21 | 313.15 | 346.80 | 2.0 | |

# Appendix E

# Threats of Death

## Estimated Risks for an American over a Fifty-Year Period

| | |
|---|---|
| Risk of death from botulism | 1 in 2,000,000 |
| Risk of death from fireworks | 1 in 1,000,000 |
| Risk of death from tornadoes | 1 in 50,000 |
| Risk of death from airplane crash | 1 in 20,000 |
| **Risk of death from asteroid impact** | **1 in 6,000** |
| Risk of death from electrocution | 1 in 5,000 |
| Risk of death from firearms accident | 1 in 2,000 |
| Risk of death from homicide | 1 in 300 |
| Risk of death from auto accident | 1 in 1,000 |

---

*Source: Tom Gehrels, ed., *Hazards Due to Comets and Asteroids* (Tucson: University of Arizona Press, 1994), as reprinted in "Top 17 Ways to Die," *Final Frontier* (March/April 1996): 23.

# Index

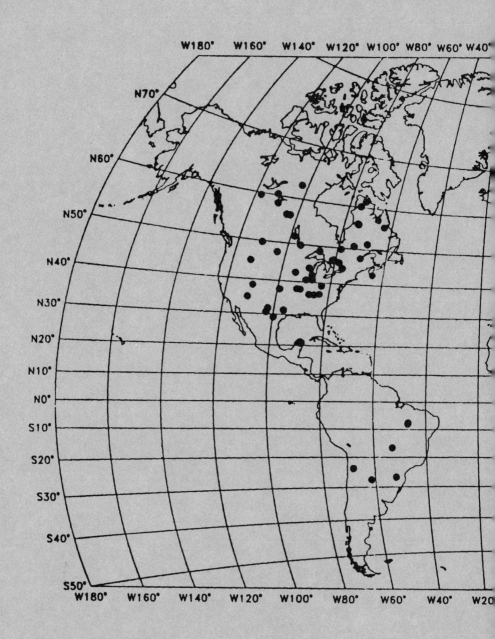